T0093964

Understanding Forensic DNA

Forensic DNA analysis plays a central role in the judicial system. A DNA sample can change the course of an investigation, with immense consequences. Because DNA typing is recognized as the epitome of forensic science, increasing public awareness in this area is vital. Through several cases, examples, and illustrations, this book explains the basic principles of forensic DNA typing, and how it integrates with law enforcement investigations and legal decisions. Written for a general readership, *Understanding Forensic DNA* explains both the power and the limitations of DNA analysis. This book dispels common misunderstandings regarding DNA analysis and shows how astounding match probabilities such as one-in-a-trillion are calculated, what they really mean, and why DNA alone never solves a case.

Suzanne Bell is Emeritus Professor of Forensic Science at West Virginia University, United States. She served on the National Commission on Forensic Science. In addition to numerous scientific publications and books, she has written multiple editions of *Forensic Science: An Introduction to Scientific and Investigative Techniques* (5th edition, CRC Press, 2019) and *Forensic Chemistry* (3rd edition, CRC Press, 2022).

John M. Butler is based at the US National Institute of Standards and Technology (NIST). He is one of the most highly cited authors in forensic science and legal medicine and an internationally recognized expert in forensic DNA typing. He is the author of the leading textbooks in forensic DNA typing and served for many years as Associate Editor of *Forensic Science International: Genetics*.

The **Understanding Life** series is for anyone wanting an engaging and concise way into a key biological topic. Offering a multidisciplinary perspective, these accessible guides address common misconceptions and misunderstandings in a thoughtful way to help stimulate debate and encourage a more in-depth understanding. Written by leading thinkers in each field, these books are for anyone wanting an expert overview that will enable clearer thinking on each topic.

Series Editor: Kostas Kampourakis http://kampourakis.com

Published titles:

Forthcoming:

Understanding Forensic DNA

SUZANNE BELL
West Virginia University

JOHN M. BUTLER
National Institute of Standards and Technology

CAMBRIDGE
UNIVERSITY PRESS

CAMBRIDGE
UNIVERSITY PRESS

University Printing House, Cambridge CB2 8BS, United Kingdom

One Liberty Plaza, 20th Floor, New York, NY 10006, USA

477 Williamstown Road, Port Melbourne, VIC 3207, Australia

314–321, 3rd Floor, Plot 3, Splendor Forum, Jasola District Centre,
New Delhi – 110025, India

103 Penang Road, #05–06/07, Visioncrest Commercial, Singapore 238467

Cambridge University Press is part of the University of Cambridge.

It furthers the University's mission by disseminating knowledge in the pursuit of
education, learning, and research at the highest international levels of excellence.

www.cambridge.org
Information on this title: www.cambridge.org/9781316517185
DOI: 10.1017/9781009043311

© National Institute of Standards and Technology and Suzanne Bell 2022 outside of the United
States of America. As a work owned by the United States Government, this Contribution is not
subject to copyright within the United States. Outside of the United States, Cambridge University
Press & Assessment is the non-exclusively licensed publisher of the Contribution.

This publication is in copyright. Subject to statutory exception
and to the provisions of relevant collective licensing agreements,
no reproduction of any part may take place without the written
permission of Cambridge University Press.

First published 2022

Printed in the United Kingdom by TJ Books Limited, Padstow Cornwall

A catalogue record for this publication is available from the British Library.

Library of Congress Cataloging-in-Publication Data
Names: Bell, Suzanne, author. | Butler, John M. (John Marshall), 1969– author.
Title: Understanding forensic DNA / Suzanne Bell, John M. Butler. Other titles: Understanding life
series
Description: Cambridge, United Kingdom ; New York, NY : Cambridge University Press, 2022. |
Series: Understanding life | Includes bibliographical references and index.
Identifiers: LCCN 2022022812 (print) | LCCN 2022022813 (ebook) | ISBN 9781316517185
(hardback) | ISBN 9781009044011 (paperback) | ISBN 9781009043311 (epub)
Subjects: MESH: Forensic Medicine – methods | Sequence Analysis, DNA – methods | DNA
Fingerprinting
Classification: LCC RA1063.4 (print) | LCC RA1063.4 (ebook) | NLM W 700 | DDC 614/.1–dc23/eng/
20220701
LC record available at https://lccn.loc.gov/2022022812
LC ebook record available at https://lccn.loc.gov/2022022813

ISBN 978-1-316-51718-5 Hardback
ISBN 978-1-009-04401-1 Paperback

Cambridge University Press has no responsibility for the persistence or accuracy of
URLs for external or third-party internet websites referred to in this publication
and does not guarantee that any content on such websites is, or will remain,
accurate or appropriate.

"This is an excellent book, clearly and comprehensively explaining the scientific aspects of DNA and its analysis and the interpretation of DNA profiles used in forensic science. The sections on emerging issues and dispelling common misunderstandings are particularly useful. The book will be an invaluable guide for scientists and non-scientists alike, and the authors are to be congratulated for an outstanding text."

Professor Niamh Nic Daéid, Director of the Leverhulme Research Centre for Forensic Science, University of Dundee, UK

"It is essential that the expectations of the public and legal professionals about forensic DNA technology are not based on crime-fiction TV shows, but rather on reality. The authors of this book explain very clearly what DNA can tell us, address existing misconceptions, and share exciting new developments. I enjoyed the real cases included in the book, all very interesting and informative. *Understanding Forensic DNA* is definitely a must-read for anyone who wants to become a DNA expert!"

Dr Lourdes Prieto, Forensic Science Institute, Santiago de Compostela; General Headquarters of the Spanish Forensic Police, Madrid, Spain

"*Understanding Forensic DNA* is the perfect resource for anyone wishing to learn about forensic DNA analysis for the very first time. It is a quick read, ideal for the legal professional and for the student who is just beginning to learn about forensic science. Unlike many books of this type, it is written from a historical perspective that outlines how the discipline has evolved in a few short decades. The book thoroughly examines all existing and past technologies in the most comprehensive way and explains the strengths and limitations of each. One would be hard pressed to think of a relevant topic that has been left out. The case examples are instructive and in no way sensational. For someone who teaches forensic science at the college and graduate level, the book provided me with fresh ideas that I can use to improve my teaching of the subject."

Lawrence Quarino, PhD, ABC-GKE, Cedar Crest College, Pennsylvania, USA

"This is a go-to book to enhance anyone's understanding of forensic DNA. It is of value to the general public, students of forensic sciences, forensic practitioners, and lawyers. It comprehensively covers all key aspects relating to forensic DNA and plenty of additional intriguing elements you may not have considered. There is a focus on the past, present, and future of forensic DNA with descriptions of key historical events. Complex issues are expertly described with helpful examples, tables, and illustrations. The authors provide a balanced view of the utilization and limitations of forensic DNA analyses, demystify common misunderstandings, and explain current concerns. Reading this book refreshed my understanding and brought new insights to my attention. Another outstanding contribution to forensic science by two exceptional authors."

Roland van Oorschot, Forensic DNA researcher

"Suzanne Bell and John Butler take a journey through the history of forensic genetics. They showcase how this exciting field emerged from its precursor sciences, hematology and serology, and guide the reader through the most recent scientific developments and debates. This entertaining trip provides an unseen density of information and is accessible to a broad audience."

Dr Walther Parson, Institute of Legal Medicine, Medical University of Innsbruck, Austria; PennState University, Pennsylvania, USA

"This book is full of information delivered in a very accessible way. It will be useful for the general public who want to inform themselves about DNA, and also for those scientists who want to be reminded of the historical basis or background of some techniques. It will be an aid to those a little daunted by the more advanced texts. The use of real-life examples throughout the text gives an excellent insight into the uses and limitations of forensic DNA. The basic facts about DNA and the techniques in routine use are understandably described in greater detail than the more modern techniques which are not so mature."

Sheila Willis, Honorary Professor, University of Dundee, Fellow of the Leverhulme Research Centre for Forensic Science, UK; retired Director General, Forensic Science Ireland

"A book that anyone with an interest in DNA profiling will really enjoy. After reading it, you will be well informed, and up to date on current developments. From the inception of biological identification methods to novel methodologies, all is well illustrated by reference to key cases and how the use of DNA profiling has changed forensic investigations. I will be recommending *Understanding Forensic DNA* as a key text to my undergraduates."

Professor Adrian Linacre, OAM, Chair in Forensic DNA Technologies, Flinders University, Adelaide, Australia

"This book has a very logical structure and touches on all aspects of forensic DNA analysis in an easy-to-understand fashion. Case examples illustrate specific challenges and make the subject matter come alive. I like that the authors include difficult concepts, for example stochastic effects, and relevant calculations, for instance for genotype frequencies and mixture ratios. Forensic DNA analysis is explained but not simplified."

Mechthild Prinz, PhD, Professor John Jay College for Criminal Justice, New York, USA

"Forensic DNA analysis contributes significantly to the provision of justice in many jurisdictions around the world. However, there are still misconceptions about what it really is. It is a technical and multidisciplinary subject encompassing biology, statistics, and the law. A book providing a clear and comprehensive description of forensic DNA analysis is very welcome, not only for public awareness but also for improving communication between those involved in this area."

Dr Roberto Puch-Solis, Forensic Statistician, Leverhulme Research Centre for Forensic Science, University of Dundee, UK

"The authors present an excellent overview of the history of forensic DNA analysis, starting with biological identification based on serology in the 1980s and moving on to the first application of DNA profiling by Alec Jeffreys to help solve a case in 1986. The chapters that follow discuss the progressive use of ever smaller fragments and quantities of DNA for identification. This evolution was accomplished by new discoveries about the human genome in conjunction with more sophisticated and sensitive analytical tools and techniques to process

DNA. The book discusses the issues that are raised when new methods are applied to DNA evidence, ranging from increased sensitivity to interpretation of data and acceptance in the courtroom. The final chapters deal with cutting-edge technology and new issues, including the use of DNA to create phenotypic profiling, familial searching, genetic genealogy, behavioral profiling, and non-human applications."

Mark Okuda, Evergreen Valley College, San Jose,
California, USA

"DNA has brought about a revolution in all fields of the life sciences, and in medicine, biotechnology, and pharmaceuticals. But, arguably, the public knows of DNA from its impact on ancestry and forensics. These are closely related, for in both cases DNA is the key to determining connections, whether between people or between evidence and perpetrator. Bell and Butler's book is therefore welcome and timely, setting out in clear and readable language the basis of DNA-based forensics and how it is used in practice. Most importantly, it also makes clear the limitations of DNA forensics, both the technical and societal issues that it raises. The authors write that their goal is 'to provide an overview of DNA methods, their use, and the current issues in the field.' They have succeeded admirably."

Professor Jan A. Witkowski, Cold Spring Harbor Laboratory,
New York, USA

"*Understanding Forensic DNA* provides an intriguing overview of one of the most innovative fields of applied sciences. Written by two compelling narrators with proven expertise in molecular and forensic genetics, this book not only covers the fascinating scientific background of current forensic DNA technologies but also delves into recent innovations such as investigative genetic genealogy that helped to identify the Golden State Killer. Similarly, other high-profile cases are used to illustrate the evidential power of a given technology. At the same time, a critical distance is maintained, such as when addressing cognitive bias or discussing the use of high-throughput genotyping that may threaten the privacy of suspects or witnesses. This book is worth reading not only for forensic science aficionados but also for

DNA experts who want to discover state-of-the-art resources for education and training."

Professor Peter M. Schneider, Institute of Legal Medicine, Cologne, Germany

"Suzanne Bell and John Butler have put together an outstanding text on forensic DNA that the general public can understand. The reader can easily appreciate the impact and limitations of the use of DNA technology through the excellent descriptions and illustrations. This book should be on the shelf of every educator, student, scientist, lawyer, investigator, or anyone working in the criminal justice system who needs to understand the role of DNA in forensic science."

Thomas A. Brettell, PhD, Professor of Chemistry, Cedar Crest College, Pennsylvania, USA

"This easy-to-understand book is a guide I will reference again and again. If you want to truly understand how DNA is used in forensic science, this is the book for you!"

CeCe Moore, Chief Genetic Genealogist, Parabon

To the unsung heroes in libraries everywhere – the research for this book would have been impossible without them. Also, to my co-author, colleague, and friend, Dr. John Butler, who has done more to make forensic science better than anyone I know.

Dr. Suzanne Bell

To Margaret Kline, my NIST colleague for many years in the DNA Technology and Applied Genetics Groups, who showed by example how to be a great scientist with her attention to detail and passion for excellence.

Dr. John M. Butler

Contents

Foreword

"I go where the evidence takes me." "Witnesses lie; the evidence does not lie." Statements like these made a strong impression on me many years ago when I watched some episodes of one of those forensic investigation TV series. The message conveyed by these statements is that forensic evidence, especially forensic DNA evidence, is the most reliable kind, and it can guide an investigator toward the person who committed a crime. However, things are not as simple and straightforward as that. As Suzanne Bell and John Butler clearly explain in this brilliant book, forensic DNA evidence can lead neither to infallible, nor to definitive, conclusions. The reason for this is that as we do not have the DNA profiles of everyone in the world, we can only figure out the probabilities that someone could have or not have the profile found at a crime scene just by chance. We also need additional evidence to confirm that the evidence found at the scene is indeed related to the crime. This does not make forensic DNA analyses useless. However, as with any science-related endeavor, forensic DNA analysis has a certain potential as well as particular limitations. Suzanne Bell and John Butler have produced the best, most accessible, and most concise guide possible to the science of forensic DNA. Some of the most striking features of the present book are the detailed explanations of the methods used, which are accompanied by very informative and clear illustrations. Read this book, and the forensic investigation TV series may never look the same again.

Kostas Kampourakis, Series Editor

Preface

Forensic DNA typing has leaped into the mainstream of criminal justice over the last few decades. DNA analysis is described as "the gold standard," and for a good reason. DNA evidence has revolutionized biological identification and given the justice system one of its most powerful tools. As with all such techniques, however, analytical power can lead to unexpected problems and consequences. The science and technology of DNA typing have outpaced general understanding, leaving policymakers scrambling to catch up. We will explore these aspects of modern forensic DNA testing methods.

Forensic DNA analysis plays a central role in the judicial system and the life-changing decisions it renders. We will introduce many such cases in the pages to come. This book is intended to explain the basic principles of forensic DNA typing and how it integrates with law enforcement investigations and legal decisions. Because DNA typing is recognized as the epitome of forensic science, increasing public awareness and understanding in this area is vital. Many misconceptions regarding DNA persist, and DNA testing is not infallible. Accordingly, we will explore the power of the technology as well as important pitfalls and limitations. A DNA sample can change the course of an investigation, with immense consequences for one or more people. The more the public understands forensic DNA typing, the more likely it will be utilized to best effect. We can accomplish this admittedly ambitious goal by clearing common misconceptions.

We will set the stage with a brief overview of biological identification, which began with blood typing techniques developed in the early 1900s. These methods led to the use of probability and statistical approaches that are still used today in DNA typing. The first application of DNA in a criminal case was in 1986 in England, in which a double murderer was brought to justice. We will see how DNA evidence, as powerful as it is, does not solve cases by itself. DNA results are always part of an extensive investigation.

Following the background information, we will move on to current DNA profiling methods and practices. What used to take days or weeks can now be completed in a few hours. Techniques have steadily improved such that DNA residues the size of the period at the end of this sentence can be typed. The ability to detect DNA transferred by touching has been a boon but also a problem. Even minute amounts of DNA, which may or may not be relevant to a case, can be detected on almost any surface. Finding someone's DNA on an item does not tell you when, how, or why it got there. Sorting through complex mixtures from "touch evidence" remains a significant challenge facing forensic DNA. The last two chapters will delve into emerging technologies and current issues, including investigative genetic genealogy.

Our goal is to provide an overview of DNA methods, their use, and the current issues in the field. This book is not intended to be a standalone textbook, nor is it a primary resource for legal or investigative purposes. If, as a reader, you wish to explore these topics further, the books mentioned in the *Acknowledgments* and the *References and Further Reading* provide a good starting point.

Acknowledgments

The foundation of this work is the three-book series by John Butler:

Butler, J. M. (2010). *Fundamentals of Forensic DNA Typing*. San Diego, CA: Academic Press/Elsevier.

Butler, J. M. (2012). *Advanced Topics in Forensic DNA Typing: Methodology*. San Diego, CA: Academic Press/Elsevier.

Butler, J. M. (2015). *Advanced Topics in Forensic DNA Typing: Interpretation*. San Diego, CA: Academic Press/Elsevier.

Selected figures and tables from these are graciously provided as non-copyrighted materials by Dr. Butler and the National Institute of Standards and Technology (NIST), US Department of Commerce.

Dr. Suzanne Bell
I owe a debt to the staff at NIST who reviewed the book and provided feedback and guidance. Also, thanks to the editors and staff of Cambridge University Press and the series editor, Kostas Kampourakis, for allowing us to add a book to the wonderful *Understanding Life* series.

Dr. John M. Butler
I appreciate the opportunity to work with Suzanne Bell on a book intended for the general public.

Points of view in this book are those of the authors and do not necessarily represent the official position or policies of the National Institute of Standards and Technology. Certain commercial equipment, instruments, and materials are identified in order to specify experimental procedures as completely as possible. In no case does such identification imply a recommendation or endorsement by the National Institute of Standards and Technology, nor does it imply that any of the materials, instruments, or equipment identified are necessarily the best available for the purpose.

1 Biological Identification

What is Identification?

Forensic DNA typing was developed to improve our ability to conclusively identify an individual and distinguish that person from all others. Current DNA profiling techniques yield incredibly rare types, but definitive identification of one and only one individual using a DNA profile remains impossible. This fact may surprise you, as there is a popular misconception that a DNA profile is unique to an individual, with the exception of identical twins. You may be the only person in the world with your DNA profile, but we cannot know this short of typing everyone. What we can do is calculate probabilities. The result of a DNA profile translates into the probability that a person selected at random will have that same profile. In most cases, this probability is astonishingly tiny. Unfortunately, this probability is easily misinterpreted, a situation we will see and discuss many times in the coming chapters.

The drive to identify individuals is as old as humanity and is not limited to forensic applications. Your signature is a form of identification, as are biomarkers such as fingerprints and facial features. Your fingerprint or face can identify you for purposes of unlocking your phone, but neither method is infallible. The same is true of DNA profiling. Any forensic identification method aims to reduce the number of people with given characteristics to the absolute minimum and express the result as a probability. Accordingly, we will approach identification through the lens of probability, as this is the best and proper way to interpret it.

In part due to advances in DNA typing methods, the concept of identification has expanded. Before the development and widespread use of DNA typing methods, biological identification was more a process of elimination than specific identification. For example, testing a bloodstain could reveal that it came from a person with type A blood, eliminating anyone with type B as a source. This finding is helpful but not definitive, as the proportion of persons worldwide with type B blood is roughly 11%, leaving a large percentage of the population as potential sources of the stain. DNA typing methods typically yield types expected to occur in one in billions or fewer. This finding is much closer to the ideal of individual identification. In later chapters, we will explore how identification in DNA terms expands to include identification of relatives and identification of ancestors.

Biological Identification

In forensic science, the goal of biological testing is to identify an individual as a possible source of biological evidence such as blood, semen, or saliva. Evidence, for example, can range from a fingerprint at a crime scene, or a bloodstain on clothing, to a discarded weapon. Testing is designed to link such evidence to a person of interest (POI) in a crime. There are other areas in which human identification is critical, including:

- Paternity testing where the identity of the father of a child is in question
- Mass disasters in which human remains (often fragmentary) need to be linked to a specific person
- Identification of military casualties and remains from current and past conflicts
- Missing person cases
- Human trafficking
- Historical investigations
- Archaeological investigations

We will touch upon all of these, but our emphasis will be on forensic applications. The methods used in biological identification and DNA profiling are similar for forensic, historical, and archaeological testing. The key differences are usually sample type and the timeframe involved. Forensic cases are contemporary or in the recent past, such as cold cases, which are unsolved criminal investigations that remain open pending discovery of new evidence. Such cases occur or have occurred in the recent past, measured in decades at most. The oldest cold cases are typically from the 1940s or 1950s, but in those cases it is rare to have testable evidence.

Archaeological cases arise from studies focusing on individuals from eras ranging from hundreds to thousands of years ago. For example, DNA has been extracted and tested from Egyptian mummies and archaic humans such as Neanderthals. Such testing relies on bones or preserved tissues and has limited capabilities compared to what can be accomplished with fresh whole blood samples.

Historical applications relate to times that fall between archaeological and contemporary eras. In Chapter 7, we explore the identification of the last Russian Tsar and his family, killed in 1918.

Biological Methods of Identification

Biological identification is based on traits that are under genetic control. This type of evidence is called *biological evidence*, and most examples are bodily fluids. Blood is the most obvious source of biological evidence. Others include saliva (oral fluid), semen (seminal fluid), vaginal fluid, urine, and feces. All are potential biological and DNA evidence sources, but they must first be located and identified as biological evidence before additional analysis can occur. The type of evidence determines what testing can be done and what information can be obtained. Terms used to describe the evaluation of biological evidence are, appropriately enough, *forensic biology* and *forensic serology*. Blood, saliva, vaginal fluid, and seminal fluid are the most exploited types of biological evidence.

Notice we described these techniques by adding the word "forensic." For use in forensic applications, these methods of identification are adapted from established biological techniques rather than independently developed by the forensic community. Blood typing for ABO groups arose from research into deaths associated with blood transfusions. The techniques were adapted for forensic use. Forensic DNA methods evolved from research in molecular biology. Advances in the field parallel, but often lag, those used in fields such as medicine, pharmacy, and genetics. We revisit this process and its consequences several times in the coming chapters.

Characterization

The first step toward exploiting biological evidence is finding it. Biological evidence can be challenging to locate and identify, particularly with small quantities on a surface containing many other materials. Suppose a reddish stain is collected from a crime scene using a moistened swab. It may appear to be blood, but it is critical to establish this identification before investing additional time and effort in the examination. There is no point in attempting further biological testing on rust, ketchup, or red paint.

Similarly, soiled bedding or underwear associated with a sexual assault can benefit from characterization before further analysis. The testing flow moves from screening tests (also referred to as *presumptive tests*) through to *confirmatory tests* if needed. Most presumptive tests target proteins characteristic of the biological fluid. If we find something that looks like blood, we first perform the presumptive test to establish that it is indeed blood before trying to extract DNA. Ideally, these tests should not have any impact on subsequent DNA testing.

Presumptive tests for blood target hemoglobin, the iron-containing protein responsible for the red color of blood. Many chemical reagents react with hemoglobin to cause a distinct color change. Among standard tests is the phenolphthalein test, which produces a pink color in the presence of hemoglobin, and luminol, in which bright light is emitted because of a chemical reaction. Hemastix, a commercial test strip used to detect blood in urine, is also employed for this task. All these tests react with small amounts of hemoglobin. They can also produce occasional false-positive (where the test incorrectly indicates a substance is present) and false-negative (where the test incorrectly indicates a substance is absent) reactions, so they are used for screening rather than definitive identification. Some laboratories conduct additional testing to confirm the result and to determine whether the blood is human. Current DNA methods are specific to humans, but the time and cost involved in the analysis are significant. Thus, performing these additional testing steps can save time and money.

Stains from semen, vaginal fluid, urine, and saliva can become visible when illuminated by alternative light sources (ALS). A typical ALS system consists of lighting sources and filters. A light source is pointed toward the surface where a stain may be present. Different light/filter combinations make residues of blood, urine, semen, and other bodily fluids easier to see. Seminal fluid contains a fluid component, called seminal plasma, and sperm cells. Test reagents detect selected enzymes found in abundance in this plasma. Another option is testing for prostate-specific antigen, or PSA (p30), using a small test device. The prostate is a small gland that sits below the bladder in males. A technique known as "Christmas tree staining," due to the colors produced, is utilized along with microscopy to find the sperm cells. The heads of the sperm cells are dyed red, and the tails green. No sperm will be present if the man has had a vasectomy, but the other screening tests will still work. Vaginal fluid is more challenging to identify as it lacks a unique protein to target. Epithelial cells from the vaginal tract are shed into vaginal fluid, and these can be detected microscopically. Finally, saliva contains high levels of a specialized digestive enzyme, amylase, which can be targeted in presumptive testing.

Successive Classification

The flow of analysis of biological evidence utilizes successive classification. Each test in a testing sequence reduces the size of the group (called a *population*) from which the sample might have come. Suppose that a red stain is found on a wall at a crime scene. It appears to be blood, so the crime-scene investigator collects it and sends it to the lab for analysis. The population from which this substance might have come includes anything that resembles dry blood, such as ketchup or red paint. The laboratory characterizes the sample, and the results suggest that it is blood, reducing the size of the potential source population. Next, the laboratory performs a species test that indicates human blood. Still a large population, but much reduced in size from the initial group. Simple blood typing shows the blood to be type A, which represents approximately 40% of people globally. In this way, successive testing allows us to reduce the number of possibilities to ever smaller numbers of potential stain sources.

Q vs. K Comparisons

Many forensic applications of biological typing involve comparing an unknown sample such as the crime-scene stain (the *evidentiary sample*) with a *reference sample* such as one obtained from a person of interest (POI). This process is referred to as a questioned (Q) sample to known (K) sample comparison. Often there are multiple references or known samples involved. Q vs. K comparisons lead to one of *three possible outcomes*. For the sake of this example, we will assume that the questioned sample is the crime-scene bloodstain from the previous paragraph, and the known is a sample collected from a POI in the case. While the Q evidence sample is characterized by screening tests, there is no need for a serological characterization of the K reference sample since it is collected directly from the POI, typically as a buccal mouth swab or a blood draw. DNA testing is then performed separately on the Q and K samples.

Suppose in the first case that the DNA profile from the crime scene (Q) was unambiguously different from that of the POI (K). This finding results in an *exclusion*, meaning that the POI could not have been the source of the crime-scene stain. The second possible outcome is an *inclusion*, which occurs if the two DNA profiles match in all respects with no unexplainable differences. Additional statistical analysis and statements would follow, as we will discuss in later chapters. Finally, test results may be *inconclusive*. This finding might arise if insufficient information, such as a partial Q DNA profile, exists to support any conclusion.

Genetics and Heredity

Human identification using biological testing rests upon genetics and hereditary control of selected characteristics. DNA profiling is possible because everyone's genetic makeup is unique except for twins arising from the same fertilized egg. As we will see in Chapter 8, tools are emerging to address this situation. Furthermore, your genetic makeup is inherited from your parents through known and predictable processes. Current DNA profiling methods do not target genes (a common misconception), but they target variable regions of DNA that follow standard rules of heredity. DNA targeted in DNA profiling comes from our cells. Figure 1.1 illustrates the key points and features.

The top frame of Figure 1.1 shows a cell with a nucleus and a structure called the mitochondrion. Both structures contain typable DNA. We consider mitochondrial DNA (mtDNA) in Chapter 7. The nucleus contains the chromosomes

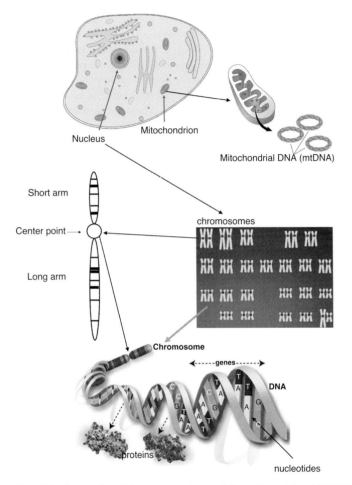

Figure 1.1   An overview of the sources and types of deoxyribonucleic acid (DNA) utilized in forensic DNA typing. DNA is found in two locations within the cell – the mitochondria (mtDNA) and the nucleus. Nuclear DNA is found in 23 pairs of chromosomes. Each chromosome is made up of strands of DNA, which organizes itself into a double-helix shape. The building blocks of DNA are nucleotides, which are illustrated in the next figure. A small portion of DNA corresponds to genes that code for proteins. The remainder of the DNA is referred to as non-coding.

(23 pairs in humans), as illustrated in the middle frame of the figure. The chromosomes have different sizes and are divided into two segments. The dark dot in the image in the middle right shows the dividing point (center point in the diagram at left). This point is essential in cell division and replication. Reproductive cells (eggs and sperm) contain 23 chromosomes, one member of each pair. These combine to form the complete chromosome set of a child. The sex-determining chromosomes are shown in the lower right of the chromosome array. Males have one X chromosome and one Y, while females are XX. DNA profiling targets these chromosomes and allows for the determination of biological sex. Another term we will use in coming chapters is *autosomal DNA*, which refers to DNA that comes from chromosomes other than the X and Y. We will also explore X and Y DNA applications for the identification and study of lineage and ancestry.

Chromosomes are made of DNA, as illustrated in the lower frame of the figure. DNA has a ladder-like configuration that is tightly folded into a double-helix shape. This shape arises from how components along the two strands bond to each other. Genes consist of long sequences of DNA. Each gene provides instructions for building a protein that has a specific function within our bodies. These proteins are large molecules capable of forming complex folded shapes, as shown in the illustration.

A closeup of the DNA structure is shown in Figure 1.2. The top frame shows the double-helix structure. The four essential compounds that link the two DNA strands are called bases – adenine (A), thymine (T), cytosine (C), and guanine (G). Their chemical structure is such that A binds with T (two bonds, as shown) while G bonds with C (with three bonds). A bonded pair such as A-T is called a *base pair*, with one base on one DNA strand and the other base on the other DNA strand. Because of the unique pair bonding of A-T and G-C, their relationship is complementary. The bonds between paired bases can be broken to allow DNA to unzip and then zip closed again in the same way since A binds to T and G to C. This ability to open and close the double-strands of DNA is central to cell replication and DNA typing.

The bases are attached to the DNA backbone, which is constructed of sugar and phosphate groups. These linked groups are the framework of the DNA molecule, with the bases facing toward the interior (the rungs of the ladder). The combination of phosphate, sugar, and base is called a *nucleotide*. One of

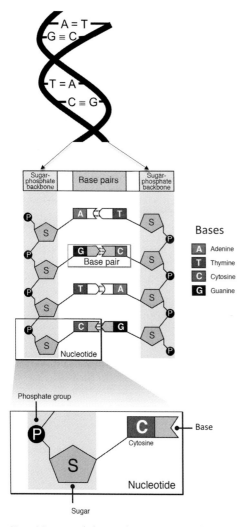

Figure 1.2   Expanded view of DNA structure. The double helix arises from a ladder-like structure with a sugar–phosphate backbone supporting rungs made of bases. The bases pair selectively (A with T and C with G) to link the strands.

the steps in DNA profiling is *amplification*, in which the existing DNA strands are copied; this is accomplished by unzipping a portion of the DNA molecule and adding nucleotides to create two copies of the DNA. We will discuss this step in more detail in Chapter 4.

Rules of Heredity

Variability in base sequences is what makes each of us biologically unique. A portion of the DNA contains information that results in protein synthesis. These proteins dictate our hereditary characteristics, such as eye color and blood type. Thus, rather than an alphabet of 26 letters, as in English, DNA information is communicated with four letters: A, T, G, and C.

Other portions of the DNA, such as those exploited in DNA profiling, do not code for proteins but still follow basic rules of heredity. Figure 1.3

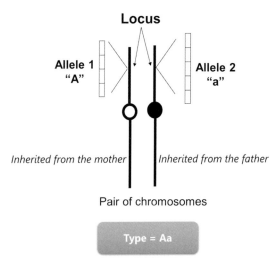

Figure 1.3   A typable location on a pair of chromosomes. The location or locus has two alleles – one from the mother and one from the father. In this example, the allele from the mother is the *A* variant and the allele from the father is the *a* variant. The person with this pattern at this locus is type *Aa*.

illustrates essential terminology. In this example, we see a pair of chromosomes with a location (*locus*) of interest, a region of the DNA targeted in DNA profiling. The copy of this region inherited from the mother and the father is shown. These variants are called *alleles*. This person has a type of **Aa** for this locus.

Additionally, this is an example of a *heterozygous type*, meaning the two alleles are different. If this person were type **AA** or **aa**, they would have a *homozygous type*. You may recall terms such as dominant and recessive referring to genes, but for our purposes we do not need to be concerned about this, as it relates to how genes are expressed.

The reproductive cells (sperm and egg) contain one copy of this chromosome and thus one copy of the DNA sequence at this locus. Now suppose a mother and a father, both with type *Aa*, have a child. The possible combinations for the child's type at this locus are shown in Table 1.1, with the mother's contribution shown in **bold** and the father's in *italics*.

The child will have one of three types – **AA**, **aa**, or **Aa**. Note that the combination of **A** and **a** occurs twice, but there is no difference between **aA** and **Aa**. The order in which the combination is written is arbitrary. Accordingly, it is possible to calculate associated probabilities for the child's type, assuming straightforward rules of heredity. Notice that the ratio of potential types for the child is 1:2:1 (**AA**:**Aa**:**aa**). Right away, you might infer that the likelihood of the child being heterozygous **Aa** is greater than that of the child being either homozygous **AA** or homozygous **aa**, since there are two ways to create this type versus only one way

| | Mother's contribution | |
Father's contribution	A	a
A	**A***A*	**a***A*
a	**A***a*	**a***a*

Table 1.1 Possible types of the child

to get **aa** or **AA**. This may or may not be the case, since we still lack critical information needed to estimate probabilities.

First, we need to know how many different alleles are found in the population. There may be only these two alleles (**A** and **a**), but there may be others such as b, B, c, C, etc. Aside from the sex-determining site, all the *loci* (plural of locus) targeted in DNA profiling have more than two alleles. Secondly, we need to know the frequency of each allele in the population. Some alleles may be common, and others may be rare; without this information, we cannot estimate the probability of a given type. We start with a simple example to illustrate this concept.

Assume that **a** and **A** are the only two alleles identified in a population and that the frequency of each is 50%. This value is commonly restated as an *allele frequency* of 0.5, meaning that half of the population will have this allele as one of the two inherited from their parents. It is important to understand that this is the allele frequency and not the *type frequency*, which we calculate based on allele frequencies. The type frequency in this example would be the number of people in the population that have the **aA** type.

In a simple case with only two alleles in a population, the frequency of one allele is assigned to a variable p and the other to q. Because there are only two alleles, we know that $p + q = 1$. This relationship is the basis of probability calculations. We can update Table 1.1 to include the p/q notation (Table 1.2).

The type probabilities can now be estimated using this relationship:

$$p^2 + (2 \times p \times q) + q^2 = 1$$

Father's contribution	Mother's contribution	
	A (**p**)	a (**q**)
A (**p**)	**A**A (p^2)	**a**A (qp)
a (**q**)	**A**a (pq)	**A**a (q^2)

Table 1.2   Possible types of the child using p/q notation

We assumed that each of the two alleles has a frequency of 0.5, so we can calculate the probability of each type in the population:

$$(0.5 \times 0.5) + (2 \times 0.5 \times 0.5) + (0.5 \times 0.5) = 1$$
$$0.25 + (2 \times 0.25) + 0.25 = 1$$
$$0.25 + 0.5 + 0.25 = 1$$

In other words, in this population, 25% of the people are expected to have type **AA**, 50% type **Aa**, and 25% type **aa**. In DNA profiling, such results are stated as a *random match probability*. What is the probability that a person selected at random from this population will have a given type? The random match probability for **aa** is the same as for **AA**, 25%, and the random match probability for type **Aa** is 50%. Alternatively, we can say that there is a one in four chance that a randomly selected person will be type **AA**, also one in four for type **aa**, and a one in two chance that a randomly selected person will be type **Aa**.

To see how allele frequency alters these probabilities, let us assume that the **a** allele is less common in the population at 20% ($q = 0.2$) and the frequency of allele **A**, the more common variant, is 80% ($p = 0.8$). We can repeat the calculation with these values to obtain:

$$0.8 \times 0.8 + 2 \times 0.8 \times 0.2 + 0.2 \times 0.2 = 1$$
$$0.64 + (2 \times 0.16) + 0.04 = 1$$
$$0.64 + 0.32 + 0.04 = 1$$

The random match probabilities are 64% for a type of **AA**, 32% for type **Aa**, and 4% for type **aa**. You can see why finding a type **aa** in a bloodstain at a crime scene would be more valuable than finding **AA**, since the **aa** type is rare compared to **AA** or **Aa**. The number of **aa** individuals is much smaller than the number of **AA** individuals, so finding **aa** eliminates many more people as possible sources than if **Aa** is found. As the number of alleles increases, these calculations become more complex, but the underlying concept is the same.

The power of DNA profiling comes from combining the probabilities of types from many loci. The calculation of *combined probabilities* is usually straightforward. The classic example is flipping a coin to see if it lands heads up or tails up. The chance of either happening is 50%, and this outcome does not depend on previous results of the coin flip. Each trial is *independent*. The probability of obtaining two heads in a row is the product of the two independent probabilities:

$$0.50 \times 0.50 = 0.25$$

There is a 25% probability (one in four) of obtaining two heads (or two tails) in a row. If the inheritance of each DNA locus is independent of the other DNA loci, the same rule (called the *product rule*) applies to DNA types and frequencies. You can continue the coin-toss calculation, multiplying the previous value by 0.5 each time, to determine that there is about a one in a thousand chance of getting 10 heads in a row and about a one in a million chance of getting 20 heads in a row.

Table 1.3 illustrates what happens when we combine frequencies, with the 50/50 chance of a coin toss as a reference. In the "common type" column, we assume that the person has the most common type at every locus. The "rare type" column represents the opposite extreme in which the person has the rarest type (here, 20% frequency) at each locus. Each table row shows the combined frequency probabilities for the given number of combined loci.

Start with the case of the common type. Combining two of these frequencies in the same way we combined coin-toss frequencies, the odds are about one in two that a person selected randomly from the population will have these two types. This value is calculated using the product rule, as we have seen before:

$$0.8 \times 0.8 = 0.64$$

To calculate the "one in" value, you can set up this relationship based on 0.64 corresponding to 64%:

$$\frac{64}{100} = \frac{1}{x}$$

You then solve for x to obtain 1.56, which can be rounded to approximately one in two.

Combining 10 frequencies of 0.8 yields one in nine, and the result is one in 87 people if 20 such loci are combined. On the other hand, if a person has the rare type at all loci, the probability of a random match is approximately one in 10 million with only 10 loci combined. Combine 20 rare types, and the value falls to around one in a trillion.

Types combined	Even type (coin toss)	Combined probability (one-in-x)	Common type	Combined probability (one-in-x)	Rare type	Combined probability (one-in-x)
1	0.5	2	0.8	1.56 or ~ 2	0.2	5
2	0.5	4	0.8	2	0.2	25
10	0.5	~ 1000	0.8	9	0.2	~ 10 million
20	0.5	~ 1 million	0.8	87	0.2	greater than a trillion

Table 1.3 Examples of combined probabilities using the product rule

These are simplified examples using two types. In DNA profiling, the only site having only two types is the site used to assess the presence of the human sex chromosomes. The two alleles are X and Y, which yield XX (female) and XY (male). All the other loci have many alleles (often ranging from 8 to 24 possibilities), producing many potential types at each locus. Because allele frequencies have been measured in various population groups, it is possible to calculate random match probabilities for the DNA profile using the product rule as outlined here. Keep in mind that the true allele frequencies are never known because the entire population is never measured. Instead, an estimate is made based on a subset of randomly selected samples in a population group. We expand on these concepts in later chapters.

Chapter Summary

Biological identification describes methods that utilize biological characteristics and features to differentiate individuals from one another. The characteristics exploited in forensic science are under genetic control, which means they follow known rules of inheritance in which one allele is from the mother and one from the father. In most situations, a forensic analysis is designed around comparisons of known (K) and questioned (Q) samples with the results described as a probability. That probability is obtained by combining individual probabilities, and the more characteristics that are used, the greater the discriminating power of the combination. In the next chapter we will examine the early methods used for biological identification in forensic science, and this will set the stage for our exploration of DNA typing starting in Chapter 3.

2 Before DNA

The Nature of Forensic Samples

Forensic samples are among the most complex encountered. Blood is best known, but other biological matrices also carry genetic information. Cheek swabs (buccal swabs) collect cells from the inside of the mouth and have the advantage of being a non-invasive sample collection compared to a blood draw. Hair, depending on the presence of the root, is amenable to DNA typing. Semen, vaginal fluids, and vaginal swabs are collected in sexual assault cases. Any surface on which biological fluids (blood, oral fluid, vaginal fluid, etc.) are deposited becomes a potential DNA source.

The initial deposition (called the *primary transfer*) occurs from a person to a surface. It is the deposition of blood, saliva, semen, or other biological substance directly from the body onto a surface. This process could be a victim's blood dripping onto an assailant's clothing, saliva on a cigarette, or seminal fluid on a bedsheet. However, this is only the first of what may be many other transfer pathways. With each transfer, the amount of material moved to the new surface decreases.

For example, assume that an assailant has attacked a victim, and during the struggle, the victim's blood is directly deposited on the knife wielded by the attacker. The assailant could wipe the blade on a towel, transferring some of the victim's blood onto the towel. This process is called a *secondary transfer*. If that towel is carried from the scene and placed on the seat of the assailant's car, a third transfer occurs. With each transfer, less biological evidence and DNA are deposited. The chain of transfers may continue, but at some point the amount of DNA remaining becomes too small to detect. These transfers can occur soon after the crime, or later, although once the sample has dried the transfer is less likely. The same transfer patterns and mechanisms are possible with other types of biological evidence.

Unlike the samples analyzed in clinical laboratories, forensic samples may be subject to deterioration and contamination. Forensic samples such as those created at a crime scene or from evidence such as clothing or bedding begin to deteriorate the moment they are deposited. Other samples collected as part of an investigation, such as reference samples collected from people in controlled conditions, are stable for months or more if they are stored correctly.

In contrast, samples created during or resulting from criminal activity may be exposed to the elements (e.g., direct sunlight, water, and heat) for hours, days, or even years in some cases. Evidence collected in older cases was often stored without regard to DNA integrity, since investigators did not know at the time what would be possible in years ahead. Evidence collected at a crime scene or autopsy, such as bloody clothing, must be thoroughly dried before storage to prevent microbial degradation.

DNA molecules can be extremely stable under certain conditions (e.g., a bone or tissue sample remaining dry and cool). Genetic typing has been performed on Egyptian mummies, and DNA sequence information has been recovered from Neanderthal bones and woolly mammoths that are thousands of years old. However, the recovery of dinosaur DNA, such as depicted in the 1993 movie *Jurassic Park*, remains in the realm of Hollywood.

As DNA methods have improved and smaller amounts of DNA can be recovered and typed, steps to prevent accidental contamination with DNA foreign to the original sample have become critical. Once in the laboratory, samples must be handled with care using gloves and proper protective garments. Special labs are isolated from the rest of the facility and have controlled ventilation like the clean rooms used in the semiconductor industry to make computer chips. DNA types of analysts and those who handle evidence are kept on file. First responders and anyone who comes through a crime scene can provide elimination samples for comparison with case results. All these safeguards are in place to ensure that DNA typing results are unambiguously associated with the evidence and the events that created it, and not from any subsequent events. DNA has been recovered from an amazing variety of surfaces, including rocks, keyboards, flip flops, toothpicks, airbags, food, stamps, envelopes, contact lenses, bullets, bullet holes, keys, and everything in between.

While DNA evidence is the focus of this book, it is vital to understand how DNA evidence came to occupy the role it does today. The use of biological evidence for identification purposes began with blood typing over a century ago and decades before the double-helix DNA structure was understood. Forensic advances paralleled those in the fields of biology, immunology, and biochemistry. The road to current DNA typing methods can be traced to the 1970s, and the first forensic DNA methods to the 1980s. Accordingly, we will begin with the ABO blood typing system.

Transfusions and Blood Types

Research related to blood transfusions led to the first method of forensic typing of blood and biological fluids. Blood transfusions had been attempted, with limited success, between animals and humans and humans and humans since the seventeenth century. Blood consists of a cellular component and the clear liquid *plasma* that can be separated by spinning at high speed in a centrifuge. Plasma is also referred to as serum. There are three types of cells (*red blood cells*, *white blood cells*, and platelets). In this book, we will focus on the red and white cells. In 1901, Karl Landsteiner from the University of Vienna in Austria discovered that when the blood serum of one person was mixed with the red blood cells of another, different outcomes were observed. In some cases the cells clumped together (agglutinated), while in other cases nothing happened. In a transfusion, cell clumping is a life-threatening event. Landsteiner realized that an immunological reaction (antigen–antibody) was the cause. Antigens and antibodies are proteins that are synthesized from genes. An excellent online resource on blood types from the American Red Cross can be found in the references for this chapter.

In short order, four blood groups were identified in humans – A, B, AB, and O. A person's blood type refers to the antigens on the surface of the red blood cell. Because there are two variants (A and B), these antigens are *polymorphic* (literally, "many forms"). A person with type O blood has red blood cells lacking antigens on the surface, while a type AB person's plasma has no antibodies. The immunological reaction that causes clumping results from antigens on the surface of the red blood cells binding to antibodies in the plasma. If the red blood cells of a person with type A blood combine with plasma from a type B person, the antibodies in the plasma bind with the antigens on the cell surface and cause agglutination. While such reactions can cause death in blood transfusions, forensic serologists exploited them to obtain blood types from evidence.

Whether or not agglutination occurs depends on the blood types of the donor and recipient. About 42% of the overall global population is type A, ≈12% type B, ≈43% type O, and ≈3% type AB. However, frequencies can vary between population groups. A person's blood type cannot identify them specifically, but it can reduce the size of the population to which they belong.

Phenotypes and Genotypes

The rules of heredity for the ABO blood group follow the same basic pattern as we noted in the last chapter, with a few additional considerations. The antigens and antibodies in the ABO system are proteins, and the loci that determine the ABO blood group are genes that encode these proteins' structures. When we refer to someone's blood type as A, B, O, or AB, this is called the *phenotype*. A phenotype is how the hereditary characteristic is observed or physically expressed. The *genotype* refers to the alleles (one from the mother and one from the father) responsible for the phenotype. In a person with an AB blood type, the phenotype and genotype are the same because they inherited an *A* gene from one parent and a *B* gene from the other. The genotype of a person with type A blood can be *AA* or *AO*. In whole blood samples, such as those tested for transfusions, the test is straightforward – plasma is separated from the cellular component of blood for compatibility testing. In an example of how science can come full circle, DNA profiling techniques can now determine ABO blood type. A person with type A blood has an A phenotype but may have a genotype of *AO* or *AA*.

Table 2.1 illustrates how the ABO type is inherited.

		Mother's genotype			
		AA or AO	BB or BO	AB	OO
Father's genotype	AA or AO	A or O	A, B, AB, or O	A, B, or AB	A or O
	BB or BO	A, B, AB, or O	B or O	A, B, or AB	B or O
	AB	A, B, or AB	A, B, or AB	A, B, or AB	A or B
	OO	A or O	B or O	A or B	O

Table 2.1 Blood types of a child based on parental genotypes

The table reveals some surprising information. For example, a mother and a father with type A blood (the phenotype) can produce a child with type O blood. This outcome could occur if both parents have the *AO* genotype. If both the egg and the sperm cells have the *O* allele, the child will have an *OO* genotype and type O blood (the phenotype).

ABO Blood Groups and Biological Identification

An early application of ABO typing for biological identification was for paternity testing. It is helpful to discuss this topic for our purposes because it illustrates inheritance patterns that apply equally well to DNA profiling. It will also show some interesting aspects of the ABO system. Consider the combination in the upper leftmost cell in Table 2.1, in which both parents have type A blood. This is the phenotype, and it means that their red blood cells have A antigens on the surface and anti-B antibodies in the plasma. Two genotypes, *AA* or *AO*, can produce a type A phenotype. A person with the *AO* genotype may have less A antigen on the red blood cell surface than an AA person, but their blood will still clump when combined with type B blood. Thus, a child of these parents can inherit an *A* or *O* allele from either parent, leading to three possible combinations (genotypes): *AA*, *AO*, or *OO*. The first two pairings yield type A blood and the last type O. In the case of one AB parent and one O parent, the child can have an *AO* or *BO* genotype and A or B type blood, but not type O.

Forensic samples are rarely whole or liquid blood. Most evidence is in the form of dried stains. As blood dries, the cells burst (lyse) and disperse the antigens. The simple clumping test with plasma and cells no longer works because the whole red blood cells have broken down. At the Institute of Forensic Medicine in Italy, Leon Lattes developed a clever method to type blood in stains. In 1915–1916, he developed the Lattes Crust test to detect antibodies indirectly. Typically, the stain is divided into three smaller pieces and placed on separate microscope slides. A few drops containing a dilute solution of red blood cells is added to each slide – one slide with type O cells, one with type A cells, and one slide with type B cells. Even though the cells in the stain have broken down, the antigens that were on the outer cell surfaces remain intact. These antigens will cause the cells to clump, which can be seen under the microscope. Suppose a stain consists of type A blood. This means that the surface of the red blood cells contains type A antigens and the serum anti-B antibodies. The slide with the dilute solution of type B cells will clump to show the stain was made with type A blood. He demonstrated the forensic value of the test in a case where he cleared an accused murderer based on an analysis of bloodstains on a coat belonging to another POI.

The other blood group you may be familiar with is the Rhesus (Rh) factor system, which Landsteiner discovered in 1937. Rh typing never developed into a common forensic typing tool despite many attempts, but it is critical for blood transfusions. A person with Rh^- type O (or O^-) blood is a universal blood donor since there are no ABO or Rh antigens in their blood. As a result, transfusions using this type rarely cause severe reactions. More than 30 other blood group systems have been identified but were never successfully adapted to forensic work. In the context of biological identification, blood groups such as ABO and Rh are referred to as *genetic marker systems*, because they are polymorphic and inherited from parents' DNA.

Secretors and Non-Secretors

Roughly 80% of us are secretors, which means that our blood group is detectable in body fluids such as saliva, mucus, and semen. Secretor status is also genetically controlled and thus a genetic marker system vital to forensic serology. If a male secretor with blood type B deposits a semen stain, the semen will contain anti-A antibodies, which can be detected using forensic techniques. Thus, if the stain is tested and shows a blood type of B, this means that the source of this semen stain is a secretor. If the investigation has identified a POI, secretor status can help differentiate between the evidence and potential sources. This ability is called *discrimination power*. Suppose, in this case, the POI is shown to be a non-secretor. He cannot be the source of this stain. More broadly, knowing that the person is a non-secretor provides discrimination power because this knowledge eliminates a substantial proportion of the population ($\approx 80\%$) as a source.

Determining blood group type was most helpful in casework for excluding or eliminating POIs. Even the smallest blood group (AB, $\approx 3\%$) represents $\approx 30,000$ people per million. Finding a type A or O sample is even less discriminating, representing more than 400,000 people per million. However, suppose blood typing analysis in a rape case demonstrated that the perpetrator was type A and a POI is developed and shown to be type B. The blood type definitively excludes him as a POI. On the other hand, if the POI in this example is type A, all this means is that he could have been the source – no more and no less. Inclusion does not imply guilt.

The combination of secretor status and ABO blood group allows combining probabilities using the product rule. Assume that 80% are secretors within a given population, and 20% are non-secretors. Further, we will assume that the ABO frequency types in this example population are 46% A, 42% O, 9% B, and 3% AB. Suppose a small seminal stain is recovered from a bed sheet as evidence in a rape case and, when analyzed, reveals an ABO type of AB. We automatically know that this person is a secretor because the stain is typable. What is the estimated random match probability for this combination? Using the product rule, we can calculate:

$$0.80 \times 0.03 = 0.024$$

This number can be expressed as one person in 42. The random match probability for the most common blood type (A) would be about one person in three. Finding the rarer type would be more useful (has greater discrimination power) because it will exclude a significant population, but alone cannot identify an individual. In contrast, finding a type A stain from a secretor has poor discrimination power because so many people are type A.

Issues with ABO Methods and a Case Example

ABO typing was widely used in forensic science for decades and provided exclusionary evidence in thousands of cases. However, problems with testing and interpretation arose in some cases. Many of the identified problems were successfully addressed using DNA typing, although at the cost of years or decades of imprisonment following wrongful convictions. The case of George Rodriguez from Houston, Texas, is one example. The offense occurred in 1987 before DNA methodology was available in US forensic laboratories.

In February of that year, a 14-year-old girl was kidnapped by two men and taken to a house where she was repeatedly raped. The kidnapping was witnessed by a 16-year-old boy who was a friend of the victim. After the assault, the victim was blindfolded and dropped along a roadside. She walked to a gas station and called the police. She was able to describe the two men and the house where the crime occurred. One of the offenders was described as fat and the other as thin.

Police quickly identified two POIs – Mr. Rodriguez (overweight) and Manuel Beltran (thin). When shown a photo lineup that included pictures of the men, the victim identified both as her attackers. The 16-year-old witness identified Mr. Rodriguez but was not able to identify the second kidnapper. Police interviewed Mr. Beltran and his brother, who implicated a different man, Isidro Yanez, as an assailant rather than George Rodriguez. Yanez and Rodriguez had similar appearances and builds. The Beltran brothers acknowledged that they had participated in the kidnapping. Subsequent investigation revealed that Mr. Rodriguez had been at work during the time the crimes occurred. He agreed to appear in a lineup and was once again identified by the victim. She identified his photo again even after noting how similar Yanez and Rodriguez appeared. This led police to arrest Rodriguez along with the other POIs.

Biological evidence, including hair, underwear, pantyhose, and a sexual assault kit, were sent to the Houston Crime Laboratory for analysis. The evidence from the sexual assault kit revealed a mixture of biological fluids from the victim and two different assailants. Table 2.2 summarizes the results.

Description	Serological results
Victim	Type O non-secretor
Evidence	Type A
Yanez	Type O non-secretor
Beltran	Type A secretor
Rodriguez	Type O non-secretor

Table 2.2   Findings in the George Rodriguez case

Since type A was detected in the evidence, at least one contributor was a secretor. Note that it is impossible to exclude any non-secretors as contributors to the evidence because they will not deposit any blood-group markers in semen. Thus, the finding that the evidence showed type A means that the *only* men that could be excluded as contributors would be type B or AB secretors. Table 2.2 shows that none of the three POIs could be excluded based on this knowledge.

The serologist concluded in her report that Beltran and Rodriguez could have contributed to the evidence. Critically, the serologist did not mention Yanez as a possible contributor, which he definitely could have been based on the ABO results. This key omission implied that Yanez was not a contributor. Adding to the confusion, the serologist reported in another document that Rodriguez was a secretor, so her finding regarding secretor status is in question.

The report's wording became critical at Rodriguez's trial, in which his defense was that Yanez was the actual second assailant. The prosecutor made a statement to the jury:

> You will hear scientific evidence which shows that beyond a doubt Isidro Yanez could not have committed the offense, and you will hear scientific evidence that there is physical evidence that is consistent with the defendant committing the offense.

This incorrect statement was based on the omission of Yanez as a possible contributor in the serology report. In testimony, the serology supervisor also stated that Yanez was excluded as a contributor, reiterating the mistake. Mr. Rodriguez was convicted and sentenced to 60 years in prison. After multiple unsuccessful appeals, hairs recovered from the evidence were analyzed in 2002 using mitochondrial DNA testing (Chapter 7). The results excluded Rodriguez, and Yanez was included. Mr. Rodriguez was released after serving 17 years.

Serum Proteins and Isoenzymes

ABO blood typing was not the only technique used for biological identification prior to DNA typing. A group of genetic marker systems using serum proteins (enzymes found in the blood plasma) was identified in the 1950s. Some of these systems are polymorphic, have hereditary types, and thus, like ABO typing, can be utilized for biological identification. The systems exploited in forensic cases are isoenzymes that catalyze biochemical reactions in the body. Isoenzymes differ in amino acid sequence but catalyze the same chemical reaction. As with the ABO system, most isoenzyme systems have only a few variants, so the discrimination power varies depending on how common a particular variant is in the population. Isoenzymes are detectable in blood and body fluids. As we saw with ABO typing, the discrimination power of most variants (A and O for example) is limited. However, when isoenzyme typing was combined with isoenzyme typing, discrimination power improved over ABO alone.

The Metropolitan Police in London was the center of forensic research and development in the middle of the twentieth century. The first system, reported in 1968 by forensic serologists Brian Wraxall and Brian Culliford, began identifying isoenzyme systems in the late 1960s. Table 2.3 summarizes the common isoenzyme systems in forensic serology.

Isoenzyme system	Abbreviation	Major types
Phosphoglucomutase	PGM	1–1, 2–1, 2–2
Erythrocyte acid phosphatase	ACP1/EAP	A, BA, B, CA, CB, C
Esterase D	ESD	1, 2–1, 2
Adenylate kinase	AK	1,2–1, 2
Glyoxalase I	GLO1	1, 2–1, 2
Adenosine deaminase	ADA	1, 2–1, 2

Table 2.3   Selected isoenzyme systems

Typing Using Gels

Culliford perfected a technique for typing these systems using a thin gel, paving the way for widespread forensic use. The procedure for typing isoenzyme systems employs gel electrophoresis (Figure 2.1). For isoenzyme systems, samples are placed into a thin layer of gel to which an electrical field is applied. These gels are hydrated porous structures usually made of a cellulose starch framework. Samples are inserted into the gel at the top in separate lanes. The electrical field is applied, and the proteins migrate through the gel, drawn to the positive terminal. The amino acids that make up the protein can acquire an electric charge based on the pH of the environment. Because the protein has a net charge and is free to move, it migrates to the opposite side of the gel. As the molecules move through the pores of the gel, large molecules move more slowly than smaller molecules. The proteins separate based on size and charge; the larger the protein, the slower the progress through the pores in the gel. The gel plate is shown as upright in the figure, but the electrophoresis process can operate vertically or horizontally. The proteins are divided into groups when the analysis is complete, as illustrated in the middle section of Figure 2.1.

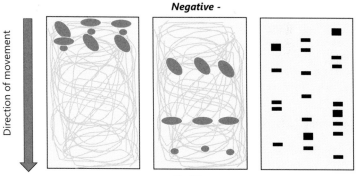

Negative -

Direction of movement

Positive +

Figure 2.1 Illustration of slab gel electrophoresis. At left, samples are loaded into wells cut into the gel at one end. When the electrical field is applied, the proteins slowly migrate through the gel matrix (middle frame), with speed and distance depending on their size and charge. Once separation is complete (right frame), proteins are visualized (e.g., by staining), revealing a banding pattern.

Proteins are not visible in the gel, so an additional step is necessary for visualization or detection. Many techniques have been used, including staining, chemical labeling, and specialized imaging. The result is a band pattern, as shown in the right-hand section of Figure 2.1. The bands are not as clean and clear as shown in the figure but are discernible. The banding pattern is the basis of assigning types. Early DNA typing produced similar band patterns.

Creating a gel plate for electrophoresis is straightforward. Dried gel flakes are boiled in water and dissolved to form a viscous liquid. The hot mixture is poured onto a glass plate with elevated sides that are a few millimeters high. After the hot gel liquid fills this mold, the excess is scraped away. As the solution cools, it thickens to a thin gelatinous slab, leading to the name *slab gel electrophoresis*. Electrophoresis is also used in DNA profiling, but it is implemented differently, as we will see in Chapter 4.

As more and better electrophoresis techniques emerged, so did arguments about which was the best. Starting shortly after the first *Star Wars* film was released in 1977, these became known as the "Starch Wars." The disputes foreshadowed what was to come with DNA and bled over into the courtroom, with experts favoring one method over another squaring off against each other in admissibility hearings. These battles took place away from public view, but privacy did not make the struggles any less vicious or painful for the scientists involved, and many of the wounds were still healing when DNA typing began to make inroads in forensic science. The era of genetic marker systems in forensic biology lasted from the late 1970s until the early 1990s. Although such protein marker systems are no longer used in criminal investigations, these innovative methods provided critical information in many cases of the era.

Combining Data

A key advantage of isoenzyme systems was additional discrimination power. By using the product rule, where probabilities from independently inherited genetic markers can be multiplied with one another, analysts combined ABO and isoenzyme types to estimate the random match probability in a given population. We worked through some examples in the last chapter, and the same principle applies here. The results were still most useful for exclusion, but more powerful than ABO typing alone. Analysts relied on population data gathered across many locations and ethnic groups. We will explore the value of combinations and population data using an example scenario. In later chapters, we see the same procedure used with DNA profiles.

The Product Rule and an Example Scenario

Suppose during a homicide the victim manages to injure the perpetrator and draw blood. The blood is deposited as drops left as the perpetrator flees the scene. During the investigation, three persons of interest are identified. One is White (POI 1), one is Black (POI 2), and one is Hispanic (POI 3). As a result, the analyst will use population frequency data for each group when using the product rule. Typing blood drawn from the three is conducted, yielding the results listed in Table 2.4 alongside example frequency data for the relevant population. For each POI, there is at least one isoenzyme type that differentiates that individual from the other two. For example, POI 1 has a different PGM type from the other two, while POI 2 has a different ACP1 type, and POI 3 has a different GLO1 type. Assuming all these types are recovered from the crime-scene evidence, it will be possible to eliminate one or more POIs based on these differences. The frequencies used for each POI are different because each POI is from a separate racial group. POI 1 frequencies are calculated based on frequency data from White populations, POI 2 from Black populations, and POI 3 from Hispanic populations. Frequencies could also be calculated from a combined frequency database. The reference frequencies used in this section were obtained from a 1987 article listed in the references for this chapter (Gaensslen *et. al.*, Isoenzymes).

System	POI 1 (frequency)	POI 2 (frequency)	POI 3 (frequency)
ABO	A (0.41)	A (0.26)	A (0.28)
PGM	1–1 (0.59)	2–1 (0.31)	2–1 (0.36)
ESD	2–1 (0.19)	1 (0.83)	1 (0.24)
AK	1 (0.93)	1 (0.98)	1 (0.97)
ACP1	B (0.40)	BA (0.32)	B (0.54)
GLO1	1 (0.19)	2 (0.48)	2–1 (0.48)

Table 2.4 Blood and isoenzyme types of the persons of interest (POIs)

This was one of the last comprehensive lists of frequencies published before DNA typing supplanted ABO and isoenzyme systems.

The table reveals several points. First, look at the AK system. Type 1 is by far the most common and thus of limited value unless a rare "2–1" type is found ($\approx$ 1–2% of the population) or the even rarer "2" type ($< 0.5\%$). We saw this same characteristic with the AB blood group. Finding a rare type is the most informative result for inclusion/exclusion.

Next, we can estimate the frequency of each of the combined types in the population using the product rule as we have in earlier examples. For POI 1, the calculation is:

Frequency $= 0.41 \times 0.59 \times 0.19 \times 0.93 \times 0.40 \times 0.19 = 0.0032 \approx 0.3\%$

This result seems to point to a tiny group, but in a town of 10,000, this would represent about 30 people. In a city of a million, the number increases to about 3000 people. Similarly, POI 2's combined frequency is ~1%, and POI 3's is ~0.6%. The field of possibilities has narrowed compared to the ABO type alone.

Now consider the crime-scene stains, which are dried blood samples. Further, assume that this sample had degraded over several days before it was collected. Typing this stain shows ABO type A, PGM 2–1, ESD 1, AK 1, and ACP1 B. The GLO1 type could not be determined, and the ABO and AK types do not provide helpful information based on common types across all POIs. We are left with the PGM, ESD, and ACP1 results, as shown in Table 2.5.

System	Stain	POI 1	POI 2	POI 3	Interpretation
PGM	*2–1*	1–1	2–1	2–1	POI 1 excluded
ESD	*1*	2–1	1	1	POI 1 excluded
ACP1	*B*	B	BA	B	POI 2 excluded
Result		Exclusion	Exclusion	**Inclusion**	
		Cannot	Cannot	Could be	
		be source	be source	source	

Table 2.5 Comparison of POIs and stain data

The PGM type eliminates POI 1 as the source of the stain, ACP1 eliminates POI 2, leaving POI 3 as the POI that could have left the stain at the crime scene. At this point, data interpretation, reporting, testimony, and wording become critical. POI 3 could be the source but cannot be identified as *the* source. Recall the calculation that showed POI 3's types were the same as ~0.6% of the relevant population (in this example, Hispanic). This value does not mean a 99.4% chance (100% – 0.6%) that POI 3 is the stain source. The 0.6% is the random match probability; if a person from the Hispanic population in this region were selected at random, 0.6% of the time they would have this same combination of types. It is also important to reiterate that the calculated percentage is an estimate based on the underlying population frequency data. The data, that the perpetrator is more likely POI 3 rather than POI 1 or POI 2, would provide investigative information and leads. The data alone does not solve the case or prove guilt.

Finally, to illustrate how frequencies within a population influence calculation, let's suppose that the three POIs had the same ABO and isoenzyme types (Table 2.6). The combined probabilities are shown in the bottom row.

Overall, the difference between ~2% and 0.6% isn't significant, but it is important to realize that different populations have different frequencies of blood and isoenzyme types. We will see a similar trend in DNA typing.

System	POI 1 White (frequency)	POI 2 Black (frequency)	POI 3 Hispanic (frequency)
ABO	A (0.41)	A (0.26)	A (0.28)
PGM	2–1 (0.35)	2–1 (0.31)	2–1 (0.36)
ESD	1 (0.77)	1 (0.83)	1 (0.24)
AK	1 (0.93)	1 (0.98)	1 (0.97)
ACP1	B (0.40)	B (0.40)	B (0.54)
GLO1	2–1 (0.52)	2–1 (0.41)	2–1 (0.48)
Combined frequencies (%)	2.1	1.1	0.6

Table 2.6 Frequencies in different populations

Population Data and Interpretation

Assume a White individual has the most common types in all the isoenzyme systems and for ABO type:

- ABO type O: 0.44
- PGM 1: 0.59
- ESD 1: 0.78
- AK 1: 0.93
- ACP1 BA: 0.41
- GLO1 2–1: 0.50

Using the product rule, this corresponds to ~3.9% of the population.

Suppose a person is selected randomly from the White population, and their blood is typed. The odds are about four in 100 or one in 25 that the types will be the same as listed, the most common types in each genetic marker system. If we select 100 people at random, we expect about four of them will have these types. This is *not* guaranteed; it is *predicted* based on the underlying frequency data. None of the 100 may have this combination of types, or more than four may have it.

Now suppose that we select 1000 people; we predict that about 40 will have these types. With this large sample set, it is unlikely that *none* would have the combination of types. If none of our 1000 people had these types, it would lead you to question how we collected the samples, our testing method, or the expected frequencies. If the frequencies in the underlying dataset are reliable and representative, the more people we select at random and type, the closer the percentage of people with all the common types will approach 4%. This is the nature of probabilistic results such as blood and DNA typing.

Two other noteworthy points. First, the population frequency dataset must be extensive and trustworthy. We do not know the true frequency of ABO types across the entire population because we have only typed a subset of all people. The same is true for subsets based on ethnic groups, sex, and geographic location. We rely on datasets to estimate the frequencies within a population, but they remain estimates. Also, databases are not static. They grow with each recorded and tabulated data entry. The more extensive the database, the more closely it will reflect the actual frequencies, which are inherently unknowable.

Finally, the product rule applies only if the types are independent of each other. The product rule no longer holds up if your ABO type exerts some genetic control over your PGM type. The product rule assumes events such as coin flips and inheritance of isoenzyme types are independent of each other. However, in some cases this is not always true. If two genes or DNA sequences are physically close to each other on a chromosome, then they may not be inherited independently. When this situation arises, it is called *linkage* and the product rule does not apply. Linkages were not an issue in forensic serological typing, but we will learn about some forms of DNA typing where linkage does matter.

Sexual Assault Cases

Blood groups and isoenzyme systems were sometimes helpful forensic tools in pre-DNA sexual assault investigations. However, sexual assault cases introduce a new level of complexity to forensic serological typing. Samples collected as part of the sexual assault kit include vaginal swabs, which will contain a mixture of contributions from the victim and the perpetrator. If neither person is a secretor, typing data will not be obtainable from the swab since neither secretes the proteins into their bodily fluids. If the victim is a non-secretor and the perpetrator is a secretor, the types obtained belong to the perpetrator only. However, since most people are secretors, the more common situation is a mix of types obtained from the swab. Interpretation of the results can be challenging, as we saw in the case of George Rodriguez. DNA profiling has dramatically improved forensic capability in sexual assault investigations, but mixtures remain a challenge, as we will see in the coming chapters.

Limitations of Blood and Isoenzyme System Typing

When ABO and isoenzyme types were used in forensic casework, the most beneficial outcome was typically exclusion. Exclusion can unambiguously eliminate a person as a contributor to, or source of, a sample. Inclusion at best reduces the number of potential sources to a fraction of the population. The need for more robust methods of typing biological evidence was clear, but had to wait for DNA profiling science and technology to become available.

Chapter Summary

ABO and later isoenzyme systems were the primary means of biological identification used in forensic science for decades. These systems follow simple rules of heredity which allowed application of the product rule to obtain combined probabilities and random match probabilities. However, forensic evidence is often difficult to analyze for these systems, given issues of sample size, degradation, contamination, and complexity (as in mixtures). Secretor status added to interpretation challenges since non-secretors were difficult to exclude in cases such as sexual assaults. It was clear to the forensic and justice communities that better methods were needed for biological identification. In the late 1980s, understanding of molecular genetics and DNA advanced sufficiently to facilitate adaptation in forensic laboratories. The next chapter focuses on the early years of forensic DNA typing. Fortunately, DNA types follow the same pattern of inheritance as ABO and isoenzyme genetic markers, so concepts we have discussed here, such as population frequencies and the product rule, apply to DNA typing as well.

3 First-Generation Forensic DNA

From Research Laboratory to Courtroom

The era of forensic DNA typing began in the 1980s when ABO and isoenzymes were the forensic tools for biological identification. As was the case with ABO blood grouping, DNA profiling was adapted from research in molecular biology. However, migration from the research laboratory to the forensic laboratory involves far more than buying new equipment. Forensic methods and techniques must satisfy two diverse communities – the scientific and the judicial. There is a common misconception that science and justice both seek "truth" and are natural partners. This assessment is oversimplified. At best, the disciplines manage to work together in a strained relationship. Before we move on to the science of DNA profiling, we need to explore how DNA found acceptance in the courts.

When a new scientific method is employed in a case, the courts must decide whether the data will be admitted into evidence that will be seen by those who will pass judgment, such as a judge or jury. The format and structure of court systems vary worldwide, but typically an admissibility hearing is used. This setting is where the clash of scientific and judicial cultures becomes apparent. The scientific community and the judicial system judge admissibility differently. Furthermore, court officials rarely have significant scientific education and experience, yet they must decide whether to admit or reject results from novel scientific methods and evidence.

Courts rely on precedent (past decisions) and guidelines (rules of evidence) to make admissibility decisions. Countries and jurisdictions have different rules and guidelines to follow, but in general, admissibility is granted when a method is deemed relevant and reliable. Factors integrated into this decision include acceptance by the scientific community, peer-reviewed publications in the scientific literature, validity of the underlying science and technology, known limitations and potential error rates, and the ability to test the methods independently. The relevant scientific community for DNA evidence encompasses molecular biology, genetics, and statistics (for data analysis and interpretation). As we move through the early DNA story, we will touch upon some of the key court decisions that moved DNA profiling from novelty to routine.

Repeating DNA fragments

The first DNA typing procedure used in criminal investigations was *restriction fragment length polymorphism* (*RFLP*). Recall that DNA molecules are composed of four different nucleotide building blocks: adenine (A), thymine (T), guanine (G), and cytosine (C). DNA methods can be designed to target inherited DNA markers that vary between individuals. The variations may be in the base sequence or the number of repeated sequences. As an example of a sequence difference, a section of DNA in one individual may be AAT**G**GCCC, while in another person it could be AAT**T**GCCC. This is an example of variation in the base sequence. The highlighted difference is in the fourth position, where the first person has a G and the second has a T. This is an example of a *single nucleotide polymorphism* (*SNP*), a topic we will explore in Chapter 8. The other type of variation occurs in the number of repeated sequences. The greater the number of repeats, the longer the DNA segment that contains it.

RFLP and current DNA profiling methods exploit this variation in length of DNA sequences due to their different numbers of nucleotide repeats. The base sequence of the repeated sequences (repeat units) is the same; what varies is their number. The number of repeats is a hereditary characteristic. For example, there is a position on chromosome 10 called D10S1248, which is used in current DNA profiling. The repeated sequence in the DNA at this locus is four nucleotides long, GGAA, and the number of repeats typically ranges from 8 to 13. For example, you might have nine repeats of this sequence inherited from your mother, that is $[GGAA]_9$, and 12 from your father, that is

[GGAA]$_{12}$. Your type at that locus would be 9,12 in that case. The loci targeted by RFLP procedures consist of longer repeated sequences called *variable number tandem repeats* (*VNTR*). If the repeated sequence is less than six or seven nucleotides long (such as the four-base sequence GGAA in the case of D10S1248), the locus is called a *short tandem repeat* (*STR*). The next chapter delves into STR protocols; here, we focus on RFLP methods with VNTR sequences which have the longer repeated units.

The repeated unit in VNTRs may range in length from about 8 to 40 base pairs (bp). Typical loci used were D1S7 (9 bp repeating unit), D2S44 (31 bp), D4S139 (31 bp), D10S28 (33 bp), D14S13 (15 bp), and D17S79 (38 bp). Notice how these loci are on different chromosomes (1, 2, 4, 10, 14, and 17, respectively). This selection is purposeful. Locations on different chromosomes are inherited independently of each other. In contrast, loci that are close to each other on the same chromosome may not be independently inherited, complicating frequency calculations. The product rule that we have discussed, and will see again, relies on independent inheritance.

Some portions of the DNA encode the instructions for protein synthesis while others, such as those used in DNA profiling, have no known function. In the past, these regions were referred to as "junk DNA," but this term is dated and misleading. While these regions do not directly code for proteins, research continues into other roles that non-coding DNA plays in the larger world of gene expression. The preferred term is non-coding DNA, which avoids the derogatory and misleading *junk* descriptor. RFLP targets loci in non-coding regions.

The RFLP Process

RFLP typing in the 1980s was a time- and labor-intensive task. Figure 3.1 shows selected RFLP steps. The process begins with the collection of a biological fluid such as blood. Next, the cell walls are ruptured to allow for the extraction of the DNA. Specialized compounds called restriction enzymes are added to the solution. Restriction enzymes locate specific nucleotide sequences and cut the DNA at those locations like molecular scissors. Gel electrophoresis separates these DNA fragments by size. Once the separation is complete, a specialized nylon membrane is pressed against

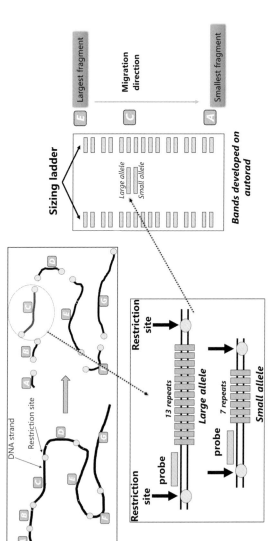

Figure 3.1   Key steps in an RFLP analysis. Restriction enzymes cut DNA into fragments (top), which are separated by size using gel electrophoresis (right). The smallest fragment (A) travels farthest while the largest stays closest to where initially placed (E). The only bands shown are for C, intermediate in size; the others are omitted for clarity. Pressing a nylon membrane onto the surface of the gel transfers the DNA to the membrane. The VNTR locations are identified by probe molecules (lower left). The probes contain radioactive atoms that emit radiation to create the band pattern on X-ray film (an autoradiogram).

the gel to capture and transfer the DNA from the gel to the membrane. Additional steps convert the DNA on the membrane to the single-stranded form, which is still bound to the nylon membrane.

A critical technological breakthrough for DNA typing methods was development of the molecular probes used to locate VNTR sites on the membrane. Probes do not bind to the VNTR itself, but to the nucleotide sequences situated on either side. Thus, the size of the DNA region that is subject to electrophoresis is larger than the repeated units alone.

The distance leading to the trailing restriction site (Figure 3.1, lower left) can be thousands of base pairs long. Radioactive atoms are included in the probe molecules. An X-ray film is pressed against the membrane, exposing the film in contact with the probes. The resulting film is called an autoradiogram ("autorad"). Reference fragments and controls (size ladders in the figure) assist in estimating sizes of specific bands.

Mutation and Mutation Rates

DNA in non-coding regions is so useful for biological identification because of differences that arise from mutation. We can all trace our ancestry back to a small group of common ancestors. Without mutations, variations in DNA would be minimal, and thus the identification power would be limited. Mutations in coding DNA may cause harm, since they may impact protein synthesis. As a result, they may be eliminated because their bearers die out. However, mutations in non-coding regions of DNA are more likely to persist, since they are not subject to evolutionary pressures. VNTRs have relatively high *mutation rates* of nearly 1%. This value means that for 100 generational pairings (mother–father), one mutation is expected in the VNTRs. We will note mutation rates for other DNA targets as they become relevant. For identification purposes, the higher the mutation rate, the more valuable a given locus will be, since this leads to more alleles, more possible genotypes, and thus greater discrimination power.

The First Cases

The first case involving DNA was not a criminal issue but a paternity and immigration one from 1985. However, the first and second cases involved the same molecular geneticist, Dr. (now Sir) Alec Jeffreys of the University of Leicester in the United Kingdom. He had discovered VNTR sequences in human DNA and realized that these were under genetic control and thus could be used as tools for identification. In 1985, Jeffreys published three landmark papers in the prestigious journal *Nature* describing his discoveries. Reflecting on his work in the early 1980s, Jeffreys stated:

> And on the morning, 5 past 9, of Monday, the 10th of September 1984, we got our first truly awful DNA fingerprint purely by chance .... We suddenly realized that we'd essentially stumbled upon a DNA-based method for biological identification. My life changed completely at that point. (Zagorski 2006)

The immigration case revolved around the youngest son of an immigrant family from Ghana. He had left the UK to visit his home country. Upon his return, he was detained because authorities suspected that he was not the son that had left the UK, but another boy attempting illegal entry. The family's attorney contacted Jeffreys to see if he could assist. The situation was complicated because the biological father was unavailable, but the mother and three other children were. Using methods that presaged familial and lineage testing (discussed in Chapter 7), Jeffreys used DNA methods to show that the boy was indeed the son who had left. The boy was allowed to re-enter the UK as a full citizen. Jeffreys later remarked that:

> If our first case had been forensic I believe it would have been challenged and the process may well have been damaged in the courts. But our first application was to save a young boy and it captured the public's sympathy and imagination. (Zagorski 2006)

The first use of DNA typing for a criminal investigation came in 1986. The DNA evidence identified the killer and cleared a suspect, which showed how DNA evidence would be used in the years to come. In 1983, Lynda Mann, a 15-year-old girl living in Leicestershire in England, was raped and murdered. Biological evidence was recovered but, without a suspect to compare to, was of no use. A second crime occurred in the same area in 1986 when Dawn Ashworth, also 15, was raped and murdered. The similarities in the crimes suggested the same man was responsible. Serological testing provided ABO and PGM types, but the combined frequency still represented about 10% of the male population. Fortunately, the attacks occurred close to where Dr. Jeffreys worked, and the police contacted him in late 1986.

The police had a suspect in custody who had confessed to the second murder but not the first. The man, Richard Buckland, was of diminished capacity, and had been seen around where the second crime occurred. Jeffreys' work showed that the DNA recovered from both scenes was the same and did not belong to the suspect. Thus, the first use of DNA in a criminal case resulted in excluding an innocent person. The police collected DNA from thousands of men in the area to find the killer, with no success. The process was christened a DNA dragnet. The break in the case came months later when a woman overheard a man in a pub describing how he had taken the blood test for another man named Colin Pitchfork. The police investigated and collected a sample from Pitchfork, which showed the same type as the evidence from both crimes. He was convicted and sentenced to life in prison.

Figure 3.2 is an autorad (i.e., the film developed following the RFLP process, as noted previously) from this case. It was instrumental in clearing Richard Buckland of the crimes. First, notice the band quality; these films were not easy to read or interpret. The two outer lanes are control samples used to analyze band position and assign alleles. Lanes 10, 11, and 12 are particularly distorted. Lanes 2 and 3 were from Buckland's blood (stain and whole, respectively). Lane 6 came from a semen stain recovered from the first victim, and lane 9 came from the sperm portion of the vaginal swab from the second victim. If you compare lanes 6 and 9, you can see the same banding pattern, which indicates that the semen from both cases came from the same man, but not Buckland. The banding pattern from his blood samples in lanes 2 and 3 is different from that found in the crime-scene samples in lanes 6 and 9. Buckland was excluded and cleared of suspicion.

Immediate Impact

The linking of two crimes using DNA evidence is something that ABO blood typing alone could not accomplish. The analysis exploited multiple loci of RFLPs, which increased the value of inclusion given the multiple loci and larger number of types compared to ABO and isoenzyme systems. Significantly, the case also illustrated the role of DNA – or any type of forensic

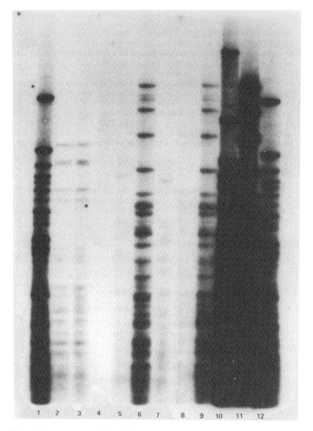

Figure 3.2 An autorad from the Colin Pitchfork double murder case. This film clearly showed the crimes were linked and that Richard Buckland was not the perpetrator. Relevant lane assignments are described in the text.

evidence. Taken alone, information from forensic evidence, even DNA evidence, cannot solve a crime. If the woman had never overheard the conversation at the pub, or if she had but never reported it, the case might never have been solved. Worse, an innocent man who falsely confessed might have been convicted and spent the rest of his life in prison. Biological evidence, even as robust as DNA, never solves the crime – investigators do.

Physical evidence is part of a more extensive investigation in which all findings must be self-consistent. We will revisit this in later sections, particularly cases that involve trace levels and mixtures of DNA. Analysis of biological evidence can yield a DNA type, but this data does not tell us anything about *how* or *why* the DNA came to be there. Finally, the case serves as a reminder that confessions must be confirmed. Richard Buckland initially confessed to the murders but was cleared when his DNA profile did not match the crime-scene profile. The real killer was identified due to police efforts to confirm (or refute) the initial confession.

The First Wave

RFLP DNA methods were rapidly adopted by many forensic laboratories and used extensively until the mid- to late 1990s. Usage by commercial laboratories outpaced that in public labs initially, which led to unanticipated difficulties. Issues arose regarding the restriction enzymes. Different restriction enzymes were used in various RFLP methods, which cut the DNA strands in different ways. This meant that resulting DNA profiles were not directly comparable among European, US, and commercial laboratories. Techniques and standards emerged to address this issue.

RFLP typing methods focused on locations showing significant variation and greater discrimination power than ABO and isoenzyme systems. However, important limitations included the lengthy analysis time and the need for a large amount of DNA for successful typing. Unlike current DNA typing methods, RFLP methods lack DNA amplification steps, so enough had to be recovered at the outset, typically more than 50 nanograms. A nanogram (ng) is one billionth of a gram. However, the DNA quantity per cell is only a few picograms (pg); a picogram is one trillionth of a gram and 1000 times smaller than a nanogram. Given the degraded condition of many forensic samples, RFLP typing often failed due to insufficient DNA quantity and quality. The method also used radioactive labels, necessitating special handling. Finally, electrophoretic separation on gels can generate distorted or smeared bands, as shown in Figure 3.2, making autorads very challenging to interpret. This last

issue, coupled with other concerns regarding data interpretation and population frequency data, generated conflicts and controversy in the courtroom and
scientific circles.

Court Rulings and Implications

We noted above that DNA was first admitted as evidence in the UK during
the Colin Pitchfork case. In the US, this milestone was reached in 1987 in
a Florida rape case. The admissibility decision was upheld upon appeal in
1988. A more serious challenge to DNA admissibility occurred
in New York City in 1989. The case involved a double murder and
a confession. In this case, a woman and her two-year-old daughter had
been stabbed and killed. A neighbor, a man named Castro, was suspected
of committing the crime. The critical evidence was a tiny drop of blood
found on his watch. A commercial laboratory analyzed the sample and
reported that the types on the watch did not match Castro but did match
one of the victims. At trial, the defense challenged the DNA results, the
methods used, and how the data was interpreted, and asked that it not be
admitted into evidence. Recognizing the critical questions raised, the court
permitted opposing experts to meet and discuss the scientific issues. The
product of the meeting was a two-page consensus report that listed issues
of concern with DNA evidence. This case marked the beginning of the
"DNA Wars," in which academic and research scientists raised significant
concerns regarding forensic procedures and practices in RFLP typing.
Castro later pleaded guilty to the charges, but the case and the associated
expert debates helped shape DNA admissibility decisions in the early
DNA era.

The following year (1990), a case in Cleveland, Ohio, dealt with similar issues
regarding admissibility (*United States vs. Yee*). Twelve experts testified in the
admissibility hearing for this case over several weeks. The FBI Laboratory had
conducted the DNA analysis. In this instance, the evidence was admitted
despite the concerns voiced by the defense experts regarding method and
interpretation.

The result of wrangling over admissibility in these and other early DNA
cases was recognition that standard procedures were needed, along with

strict quality assurance and quality control. These measures were essential before RFLP evidence would be routinely accepted. In the US, the FBI was already moving in that direction. The National Research Council (NRC), a body representing the National Academy of Sciences in the US, worked on a comprehensive DNA evidence report. The first report, *DNA Technology in Forensic Science*, was published in 1992. Many of its conclusions and recommendations were not well received, leading to a second NRC effort.

The Evaluation of Forensic DNA Evidence, referred to as NRC II, followed in 1996. One of the critical points addressed was how to define population groups. As with ABO and isoenzyme types, frequencies can vary between groups. Suppose a suspect is Hispanic. The frequency of their types should be evaluated relative to population data from Hispanics. However, a person of Hispanic origin may come from Mexico, Puerto Rico, the Caribbean, other regions in Latin American, or belong to a second or a third generation living in the US. These groups are examples of substructures in populations. Because of the challenges of defining populations, the first NRC report recommended using the highest frequency for any group as a *ceiling*. For example, if a type is found in 10% of one subpopulation, 12% in another, and 15% in another, the ceiling or largest value (15% here) would be used. In other words, suppose the population of a city is 10,000 people. If we use the 10% value, this means 1000 people are predicted to have this type. Using the largest 15% value, 1500 people are predicted to have this type. The larger the number, the more people are included as possible sources. This is considered a conservative practice because the smaller the pool of potential contributors, the greater the discrimination power of the frequency. This practice gave the benefit of any doubt to persons of interest, but often decreased the weight and utility of the evidence by being overly conservative.

The disadvantage of this approach is that it may artificially decrease the weight of the evidence. Discomfort with the ceiling approach spurred many population studies that added to the available frequency data in different groups. As we noted before, the quality and size of the underlying frequency databases are a critical and often unappreciated aspect of forensic DNA typing.

Other Key Cases

RFLP typing was used in many high-profile cases, including a 1998 analysis of Monica Lewinsky's famous blue dress. The FBI analysis showed that the semen stain from the dress was attributed to President Bill Clinton. Seven VNTR loci were typed, resulting in a random match probability of 1 in 7.8 trillion in the US White population. While this case was salacious and scandalous, the most notorious was that of O. J. Simpson, a few years earlier. The notoriety of the case, trial, results, and media coverage impacted the justice system and forensic science in many unanticipated ways.

Over 100 pieces of biological evidence consisting primarily of blood droplets and stains were gathered from the crime scene where O. J. Simpson's ex-wife Nicole Brown and her friend Ron Goldman were found dead. DNA samples were sent to three laboratories for testing using both RFLP and newer techniques (covered in the next section). Simpson's defense team vigorously attacked the collection of the biological material from the crime scene – and through accusations of improper sample collection and handling as well as police conspiracies and laboratory contamination, the defense team managed to introduce a degree of "reasonable doubt." After a lengthy and exhausting trial, the jury acquitted O. J. Simpson in October 1995.

Since then, forensic DNA laboratories have improved their vigilance in conducting DNA evidence collection. The issuance of the FBI Quality Assurance Standards, first in 1998 with updates in 2009, 2011, and 2020, has raised the professional status of forensic DNA testing in the United States. Thus, the O. J. Simpson trial helped raise public awareness of DNA testing and renewed emphasis on the importance of careful DNA evidence collection and quality assurance efforts in forensic laboratories.

PCR

Limitations of RFLP include the lengthy processing time, the amount of DNA required, and problems with degraded forensic samples. Amplifying the amount of DNA in a sample remedies the first two of these issues by increasing efficiency and the quantity of DNA available to work with. The *polymerase chain reaction* (*PCR*) accomplished this and revolutionized

DNA typing. The PCR process was first published in 1985 by Dr. Kary Mullis and colleagues at Cetus Corporation. The work garnered a Nobel Prize in Chemistry for Mullis only a few years later, in 1993. Like many landmark ideas, it was a classic "Why didn't I think of that?" kind of discovery that, in retrospect, seems obvious. In simplest terms, the PCR process involves unzipping the double-stranded DNA molecule, exploiting the fact that bases are complementary to make a copy of each strand, and zipping the two new strands closed. With each cycle of this process, the number of DNA molecules doubles. Billions of copies can be created in about 30 PCR cycles.

Extracted DNA is placed in a small vial with a heat-tolerant enzyme and ingredients needed to build new strands. Short DNA molecules (primers) are used to find and bind on either side of the region of DNA to be copied. PCR utilizes thermal cycling for opening and closing the double DNA strands. First, the sample is incubated to 94 °C for a short time. Heating a DNA molecule to almost boiling temperatures (water boils at 100 °C) causes the two complementary strands to separate or *denature*. Informally, this process is referred to as unzipping the DNA. Next, the solution is rapidly cooled to 60 °C. The primers bind (anneal) to target sequences surrounding the region to be copied. The solution is then heated to 72 °C, where the enzymes needed to make the new strands operate best. Nucleotides in the solution bind with sites on the unzipped portion based on complementarity, A to T and C to G. The new units are incorporated like building blocks by the enzyme to create a new double-stranded DNA section. The region copied is the section of DNA which lies between the two primer sites. The amount of DNA doubles with each PCR cycle. Assuming that the copying process is 100% efficient, a single DNA template molecule generates 256 copies at the end of the 10th cycle, 262,144 at the end of the 20th cycle, approximately a billion copies after 32, and more than 4 billion after 34 cycles. These numbers are for one DNA template molecule in a sample that typically contains tens, hundreds, or more templates. This PCR process can be completed in a few hours or less, depending on the time to cycle between temperatures.

PCR techniques evolved during the same period as RFLP typing was being implemented in forensic laboratories. There were attempts to use it in that process but with limited success, primarily due to the VNTR repeat region's size targeted for copying. The advantages of PCR in forensic applications were evident, which led to a search for loci more amenable to amplification. In general, shorter sequences are better suited to copying than larger ones, so the hunt was on for highly variable shorter sequences. The first of these were developed and marketed in the early 1990s.

Initial PCR Typing Techniques

The first PCR-amplified locus used in forensic laboratories was the DQα (also called DQA1) system, which is related to antigens found on the surface of the white blood cell (compared to the ABO antigens found on the red blood cell). With the original tests, the DQα locus had six detectable alleles (Chapter 1, Figure 1.3) and 21 detectable genotypes. Cetus Corporation (where Kary Mullis was employed) manufactured and marketed the first commercial kit in 1990. This marked the beginning of what would become, for better or worse, a situation where suppliers influence and control which genetic loci are tested and how. We revisit this topic in the next chapter. The Cetus kit was a test strip design that produced color changes, which is much simpler to interpret than gels and films. An expanded kit followed called Polymarker that targeted DQα (types and subtypes) along with five other DNA loci. The added loci were not as variable as DQα, with two or three types per marker.

A few other typing methods were introduced in the 1990s, but none found widespread use in forensic applications. The larger the DNA sequence, the more it is subject to degradation and decay in forensic samples. Thus, by the late 1990s, it was clear that advances in forensic DNA applications required smaller repeat sequences. It was also clear that PCR amplification was essential, because so many cases involved small amounts of DNA. In addition, better tools were needed to detect and interpret DNA mixtures in sexual assault cases. Finally, the methods had to be amenable to automation. These requirements pushed the technology forward. By the mid-1990s, all the pieces were in place for these advances to coalesce and cross into forensic application, leading to the DNA typing techniques used today. The inheritance and product rules still applied, but DNA profiling was about to take a significant leap forward.

Chapter Summary

DNA typing methods target regions in the molecule in which base pair sequences are repeated. The early methods (RFLP) focused on larger repeating sequences (VNTRs) that were first applied in 1986 in the Colin Pitchfork case. RFLP was used in other high-profile cases and quickly supplanted ABO and serum protein typing as the preferred method for biological identification in forensic laboratories. Implementation of PCR to amplify DNA samples was a key development in this period, allowing for the analysis of much smaller stains than was possible with serological and early RFLP techniques. As revolutionary as these methods were, the combination of PCR with shorter repeating units would soon transform DNA typing into the standard methods that are in use today. The next chapter will show how that happened, and how the methods work.

4 STR Methods and Loci

What are STRs?

In the last chapter, we discussed the first method (RFLP) used in DNA typing. This procedure targeted relatively long DNA fragments (VNTRs) containing many repeated units of a base pair sequence. We ended by noting that they were not amenable to automation and therefore not destined for widespread forensic applications. However, by the early 1990s, many factors coalesced to set the stage for a leap in DNA typing capabilities. For example, the forensic and legal community had adjusted to DNA evidence, and analysts had moved from serological techniques such as ABO to genetic typing utilizing multiple DNA markers. Additionally, researchers in molecular biology, including in genomic sequencing, had identified many shorter repeat sequences that exhibited variation among individuals in a population. These segments, called *short tandem repeats* (*STRs*), contain repeated sequences, such as GATA GATA GATA or ACT ACT ACT ACT, with a range of different allele lengths due to varying numbers of repeat units (e.g., three GATA or four ACT). The shorter size of these DNA regions was advantageous for many reasons, including improved success with the environmentally compromised samples commonly found in forensic casework. Thus, the forensic community began to move away from the RFLP methods we discussed in the last chapter.

Multiple advancements facilitated the forensic adoption of STR typing procedures, including an instrumental method of electrophoresis, known as *capillary electrophoresis* (*CE*), which replaced slab gel methods and allowed automation of the separation step. We will discuss this technique in more detail shortly. Multiplex PCR primer sets for simultaneous amplification of multiple markers became commercially available at about the same time. These STR kits significantly reduced the time and labor compared to RFLP. Migration to STR methods moved quickly in the late 1990s and early 2000s.

Thousands of STR regions have been discovered in the human genome with repetitive sequences containing 2–7 nucleotides. Most forensic STRs are four-base sequences (e.g., GATA or TCAT). One of the first tasks was to determine which loci to use in forensic applications. Many factors were evaluated, including the size of the repeated sequence, the overall size of the region in which the repeat was located, whether the sequence was easily incorporated in methods targeting more than one site, and how well the amplified DNA separated with capillary electrophoresis. Once standard STR loci were selected, kits could be manufactured targeting those loci and facilitating rapid transition in forensic laboratories.

All of these rapid advances in DNA typing methods and capabilities facilitated the establishment of databases, discussed in the next chapter. The United Kingdom initially targeted 6 STRs in 1995, then expanded to 10, and currently uses 17 markers. In the US, the FBI selected 13 loci in 1997, which grew to 20 core (required) loci in 2017. In 2009, the EU adopted a standard set of 12 loci. Interpol specifies 8. These various core sets of markers contain some STR loci common to all the lists, but a few differences mean that no one international standard set exists. As a result, searching outside a national database is not straightforward. Figure 4.1 shows STR loci and their locations. The dotted line is the center point, or centromere, of each chromosome, as was noted in Chapter 1 (Figure 1.3). Notice how loci are spread among several chromosomes, as was the case with RFLP loci. Chromosomes 2, 5, and 12 have more than one locus, but these are far apart on the chromosome. As a result, STR types are inherited independently, and the product rule applies.

Identification of Biological Sex

The biological sex of the person contributing to the sample can be determined along with STR types. In forensic DNA analysis, biological sex is defined as: females possess two X chromosomes, and males have one X and one Y chromosome. The X and Y chromosomes shown in Figure 4.1 contain the *amelogenin* (*AMEL*) site used for sex typing. AMEL is unique; it is not an STR

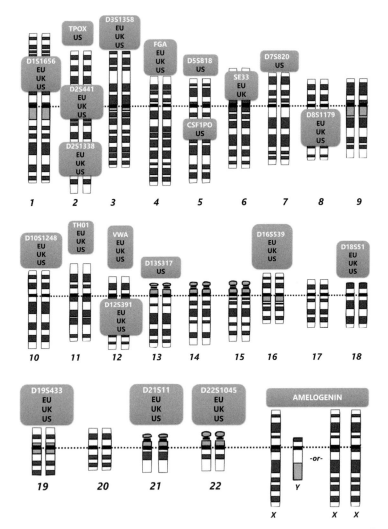

Figure 4.1 Locations of commonly typed STR loci by chromosome number. The name of each is shown in the gray boxes, along with where they are used.

marker. Instead, it is a region within the amelogenin gene. This region is six nucleotides smaller on the X chromosome than on the Y chromosome, and it can be amplified using the same approach as used for STRs. A male sample will produce two products for its X and Y chromosome, while a female sample yields only one product. With the AMEL X chromosome allele, a female sample (XX) typically has twice the amount of AMEL as a male sample (XY). Combinations such as XXY and XXX exist but these are rare.

STR Typing

Current methods of STR typing (also called *DNA profiling*) integrate steps we have seen before, with minor modifications. Among the critical technical advances that drove the rapid adoption of STR typing was development of an instrument capable of analyzing tens to hundreds of samples a day depending on the instrument configuration. We noted in the last chapter that early methods utilized slab gel electrophoresis. It worked, but it was slow and had limited sample capacity. Commercially available capillary electrophoresis instrumentation changed the game. The principles are the same as in slab gel electrophoresis, but the DNA fragments separate in a capillary tube filled with a gel-like polymer rather than across a gel bed. The capillary is about the diameter of 10 head hairs. Samples separate in the capillary tube and are detected as they pass a detection window. The bands in the gel (see Figures 2.1, 3.1, and 3.2) become peaks in a plot called an *electropherogram*. The concept is illustrated in Figure 4.2. The top frame is a reproduction of Figure 2.1 showing how samples separate in slab gel electrophoresis. The middle frame shows an example of DNA fragments separated using this technique. The lower frame illustrates how an analogous separation appears when conducted using capillary electrophoresis. The circled pairs of bands would appear as paired peaks in the electropherogram.

DNA fragment separation in CE is based on the same principles as utilized in slab gel electrophoresis. Figure 4.3 illustrates the concept and shows the interior of a typical instrument. Samples are introduced into the inlet and drawn into the tube, which acts like a miniature gel bed. As each component separates and passes in front of the detection window, a laser beam illuminates it, causing dye compounds to fluoresce. We address how and why the samples fluoresce

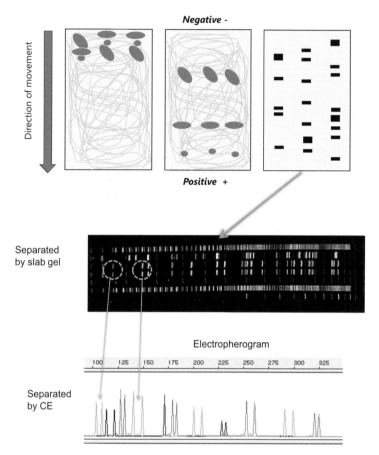

Figure 4.2   Comparison of a slab gel separation (top and middle frames) to capillary electrophoresis (CE, bottom frame). Note for example the pair of bands in the gel (circled) which correspond to peaks when separated by CE. See the text for a detailed description.

shortly. The output is a series of peaks instead of colored bands in a slab of gel. Instruments come with multiple capillary tubes and sample trays that can run continuously through a set of samples. Multi-capillary genetic analyzer CE

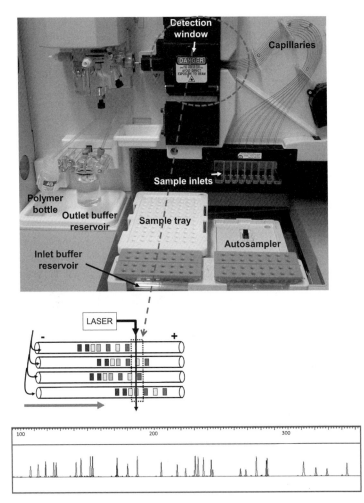

Figure 4.3   The top frame shows the interior of a commercial DNA analyzer. The capillary array is at the upper right. The capillaries direct flow into the detection window, where a laser illuminates them and causes the dyes to fluoresce for detection. The lower frame depicts capillary tubes with the separated fragments flowing from left to right. The sample flows through the detector and into the outlet reservoir. The resulting output (the electropherogram) is shown in the lower frame. We will describe this in detail shortly.

instruments are available with 4, 8, 16, or 24 capillaries and capable of processing up to two trays with 96 or 384 samples. Some high-throughput versions use arrays of 48 or 96 capillaries. Next, we will explore each step in an STR analysis, including instrumental analysis.

Steps in STR Typing: Collection and Characterization

DNA samples can come from several sources, ranging from pristine, freshly drawn blood samples (referred to as *reference samples*) to minute specks of degraded body fluids (as in forensic casework samples). Collectively these are referred to as biological evidence. Because DNA is found in almost all cells (except red blood cells), any biological material that contains cells or cellular debris is a potential source of DNA evidence. However, not all biological materials found are relevant to the case at hand. Thus, selecting and collecting evidence is a critical and often underappreciated aspect of DNA typing.

A crime-scene investigation generates crime-scene samples and *elimination samples*. Elimination samples are collected to eliminate profiles that are not associated with the criminal act. Assume a crime takes place in a home. Police officers, first responders, and crime-scene personnel responding to the scene inevitably shed biological material while there. Whoever lives in the house or has legitimate reasons to be there are sources of biological material that are not relevant to the crime. Such biological evidence is part of the natural background at the scene. Elimination samples ensure that these legitimate sources are accounted for in the subsequent analysis and interpretation.

Whenever a sample is collected, it is immediately subject to strict protocols to ensure validity and safety. A *chain of custody* document accomplishes this task. If a crime-scene analyst decides to collect a towel that appears blood-stained, the analyst first documents the evidence with digital imaging before it is touched or moved. An identifier is assigned to the item, and the chain of custody begins. In this example, the towel would be placed in a paper bag and sealed with evidence tape. Plastic bags are not used for this type of evidence as airflow is needed to allow the stains to dry. Every time the evidence changes hands, the chain of custody is updated. The transfer date and time are indicated, along with names and signatures. The document clearly shows who is or was responsible for a piece of evidence from the moment of collection.

Breaks in the chain raise questions regarding the security of the evidence and any data obtained from it.

Reference and elimination samples are collected by *swabbing*. The most common method is cheek swabbing, in which a sterile dry cotton swab is rubbed gently inside the cheek to collect cells. The swab is placed in a protective sleeve for transport and storage. A tool that looks like a small toothbrush can also be employed. Swabbing is also used to collect biological evidence from a scene or surfaces such as clothing. Double swabbing can be utilized with dry materials. A long cotton swab is dipped in sterile water to moisten the surface, which can help rehydrate cells. A second dry swab is then brushed across the surface to collect these residues. Both swabs are retained and analyzed. Cellular material can be collected from dry surfaces using tape as well. Any damp evidence must be dried before storage to minimize mold growth and degradation. Ideally, dried items are stored in temperature-controlled rooms or under refrigeration (4 °C) or even in freezers (–20 °C). Before DNA testing, evidence samples are usually characterized with screening tests, as we discussed in Chapter 1.

Steps in STR Typing: Extraction

Once samples are characterized and selected for typing, the next step is DNA *extraction*. Extraction has two goals – to remove enough DNA from the matrix (like a swab) for analysis and to isolate it from other biological materials associated with cells, such as proteins. DNA must be released from cells, which requires breaking the cells open (*lysing*). Extractions do not always purify the DNA completely. Inevitably, other compounds are extracted along with DNA, and these may interfere with the DNA analysis procedure. We discuss such issues further in Chapter 6. Cellular components like proteins may also interfere with (*inhibit*) reagents used in PCR. Extraordinary care is taken in the laboratory to prevent accidental contamination. For example, crime-scene samples may be prepared in a separate laboratory area from reference samples to ensure no unintentional mixing of the two.

The amount of DNA present in samples varies. Table 4.1 provides a summary of DNA content typical of forensic samples. For reference, a regular metal paper clip weighs about 1 gram; a milligram (mg) is a thousandth of a gram;

Type of sample	Amount of DNA
Liquid blood	20,000–40,000 ng/mL
Bloodstain	250–500 ng/mL
Liquid semen	150,000–300,000 ng/mL
Postcoital vaginal swab	10–3000 ng/swab
Plucked hair with root (contains cells)	1–750 ng/root
Shed hair with root	1–10 ng/root
Liquid saliva	1000–10,000 ng/mL
Oral swab (buccal swab)	1–20 ng/mL
Bone	3–10 ng/mg
Tissue	50–500 ng/mg

Table 4.1   Typical DNA amounts found in biological evidence

a microgram (μg) is a thousandth of a milligram; and a nanogram (ng) is a thousandth of a microgram. One nanogram therefore corresponds roughly to a paper clip that has been cut into a billion pieces. A milliliter (mL) equals about 20 drops from an eyedropper. It is also the same as a cubic centimeter (cc), a unit widely used in medicine. DNA originating from solid materials (bone and tissue) is given in units of weight (ng/mg) in Table 4.1.

Forensic laboratories have several options for sample extraction. The first step is to place the sample (such as the top of a cotton swab) in a solution that contains extraction and treatment reagents. The container used is about the size of the tip of a pinky finger. Possible additional steps include the use of other reagents and centrifuging to separate solids from liquids. After centrifuging, solids are packed tightly into the bottom of the vial with the remaining liquid (the supernatant) above it. The DNA is in the supernatant, which is drawn off from the solid residue.

Sexual assaults generate mixed samples, most often male–female. This scenario calls for a different approach in which sperm cells from the perpetrator are separated from the woman's vaginal *epithelial* cells (cells found on the surface of the vaginal tract). The first step is to add the extraction solution, which contains the ingredient required to lyse (break open) the epithelial cells. The female DNA remains in the solution, while the sperm cells can be isolated by centrifuging. The supernatant is removed and set aside as the female fraction.

The sperm cells are lysed with a different set of ingredients to release the male DNA from the male fraction. This differential extraction procedure is possible with sexual assault samples because of a strong protective cell coating on intact sperm cells.

Steps in STR Typing: Quantitation

Extractions isolate DNA – any DNA- from the original evidence. Samples from crime scenes may be contaminated with DNA from other sources such as bacterial, animal, or plant material. The quantitation step measures how much of the recovered DNA is human. In addition, most forensic DNA tests are designed to work within a narrow range of DNA amounts, $\approx$ 0.5–2.0 ng. Too much DNA can present as many problems as too little, so knowing the amount is essential before the PCR amplification step. If too much DNA is present in the extract solution, dilution can solve the problem. If too little is present, evaporating off some of the liquid to increase DNA concentration is an option. Several quantitation methods are available in forensic laboratories.

Steps in STR Typing: Amplification

Recall that the amplification process begins with the unzipping of the DNA double strands by gentle heating. Once the strands are separated, *primers* locate and surround the region to be amplified (copied). Rather than finding the actual repeating sequence, the primer locates the *flanking areas* surrounding the STR site. The base pair sequence of the two flanking regions (referred to as forward and reverse) are not the same, so two primers are needed for each STR. One of these two primers is labeled with a dye that fluoresces when illuminated by a laser. This light signal is what is detected by the capillary electrophoresis instrument.

The example in Figure 4.4 shows an STR with GATA as the repeated sequence. In this example, the maternal STR has six repeats, and the paternal has eight repeats, yielding a 6,8 genotype. Genotypes are typically listed with the smaller allele first. The PCR process copies the entire region from the start of the forward primer to the end of the reverse primer, so the fragment amplified is bigger than just the repeating units. We saw the same thing with RFLP,

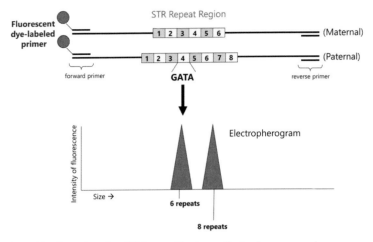

Figure 4.4 Illustration of an STR locus with primers. Electropherogram peaks are represented by the triangles. In this example, the forward primer contains the fluorescent dye used in detection. The region that is amplified during PCR stretches from primer to primer. The repeated unit in this case is GATA. The chromosome from the mother (maternal) has six repeats and the one from the father has eight, so this person has a 6,8 genotype. The electropherogram shows two peaks, one for each allele.

where portions of the DNA flanking the repeat region were also part of the DNA fragment measured. When the amplified sample is analyzed using capillary electrophoresis, the peak pattern for this sample at this locus would be as shown in Figure 4.4, lower frame. Shorter DNA fragments (e.g., allele 6 compared to allele 8) move faster in the capillary tube and emerge first. As each passes the detection window, the laser causes the dye to fluoresce, producing the plotted signal in the electropherogram.

Companies and Kits

The materials needed for amplification of an STR marker are provided as a commercial kit. Laboratories typically purchase multiplexing kits that target the STR loci being typed. The most critical component of a kit is the primer mixture. Other components include:

1   A PCR buffer solution for the reaction
2   Nucleotides of A, T, C, and G to construct copies
3   The polymerase enzyme used to extend the DNA strands as they are being copied
4   An allelic ladder for each locus
5   Positive control samples to demonstrate the kit is working properly

The allelic ladder for each STR marker is a mixture of all its common alleles. Figure 4.5 shows the ladders for each STR marker targeted by this particular commercial kit. The color label at the upper right of each row describes the color of the fluorescent dye associated with the labeled primer. Another essential

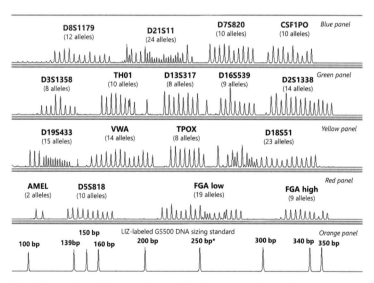

Figure 4.5   An electropherogram showing alleles for 16 STR markers and the AMEL site. The lowest row shows peaks obtained from DNA fragments of known sizes, which help to identify the peaks in the rows above. The rows are divided and grouped by the color of the dye in the corresponding primers. These peaks appear in that color on the computer screen.

feature is the internal size standard shown in the bottom row. This plot shows the relative sizes of the amplified STRs. As one example, the sex-determining AMEL marker appears at the left of the fourth row down. It has two alleles and is flagged by a dye that appears bright red when it fluoresces under laser illumination; a male source yields two peaks and a female yields one. It is the smallest amplified fragment labeled with the red dye in this kit configuration. The third row down shows D19S433, an STR locus on chromosome 19 with 15 alleles labeled with a yellow dye.

Kit manufacturers may utilize different primer sequences for each STR locus. Kits containing materials for more than one STR locus are called *multiplexes*. Performing the research to identify these primer sequences requires a significant investment of time and resources by the manufacturer, which is one factor that contributes to the cost of commercial STR typing kits. In the last chapter, we discussed commercial kits manufactured for RFLP and DQα typing methods. The same pattern continues into STR typing. Forensic laboratories can create primers, kits, and assays, but this requires significant effort and thorough validation. Companies provide forensic labs with a vital product and published validation information, but as a result, manufacturers determine which STR loci are typed and available.

Legal issues arose in the early days of STR methods regarding commercial primer sequences. Kit manufacturers were reluctant to release this information as it would provide competitors and laboratories with the critical information needed to create their own primers. With primer sequences considered proprietary, courts were initially reluctant to admit STR typing without this information. One company (Promega) published its STR kit primer sequences and obtained patents to protect aspects of its multiplexing technology. Other manufacturers elected not to publish their primer information but instead published results from their validation studies. We will see other conflicts regarding propriety information and software in later chapters.

Steps in STR Typing: Typing and Interpretation

The goal of the STR process is to identify the genotypes found in the sample. Figure 4.6 illustrates this process. Data from the capillary electrophoresis instrument appears as a series of peaks as shown in the top

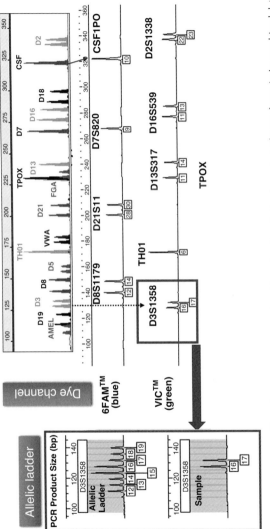

Figure 4.6 Illustration of how STR genotypes are assigned based on the electropherogram. The top frame is the combined data, with the example locus D3S1358 highlighted. The dye color used with the primer for this STR is green, so the data is displayed in the green channel. The allelic ladder for this locus is shown at the left. The scale on the top refers to the number of base pairs in the DNA, which is obtained from the size ladder (shown in Figure 4.5). The sample peaks are identified by aligning the sample with the allelic ladder.

frame. The displayed color corresponds to the dye label attached to the primer. The instrument display can be viewed as separate dye channels in which all the blue-labeled STR peaks (first frame) and all the green-labeled peaks (second frame) are grouped. The yellow, red, and orange channels are not shown here for clarity. The STR loci are identified by color and size, and their names appear above the peaks, as illustrated in Figure 4.6. Although a few tri-allelic patterns exist, each STR site should have either one or two peaks in a single-source sample. If the person is heterozygous at this site (the mother and father had different repeat numbers), two peaks appear. If the person is homozygous (the same number of repeats from each parent), one peak appears. The person who contributed this sample is homozygous at loci D7S820, CSF1PO, and TH01.

The number of repeats represented by each peak is determined using the allelic ladder. The ladder for D3S1538 appears at the left of Figure 4.6. Alignment of sample peaks with the ladder indicates the number of repeats and the resulting genotype. This person is type 16,17 at this locus. This process is repeated for all the STR locations to generate the person's full STR profile from all the sets of genotypes observed. The product rule is used to calculate the probability of this set of genotypes occurring at random. For the individual whose types appear in Figure 4.6, this probability is one in quintillions. This finding drives home the power of DNA typing using STRs over older methods of biological identification. Even the rarest ABO blood type occurs in ~2% of the population.

Trusting the Method

DNA typing using STRs has become routine in forensic laboratories. However, the discussion above does not touch upon all the effort and labor behind a single DNA analysis. Multiple layered procedures ensure the quality, utility, and reliability of the data produced. Collectively these practices are referred to as *quality assurance and quality control* (*QA/QC*), and they start long before any samples reach the forensic laboratory. Quality assurance includes systematic practices and procedures, while quality control refers to the day-to-day methods used to ensure that results meet strict quality standards. Analysis of the positive controls supplied with kits is an example of a quality control step. Ensuring analysts are trained to the current methods and having a second analyst review the data are examples of quality assurance. The quality system goes beyond the laboratory walls and encompasses the international community of DNA analysts. Table 4.2 lists examples of quality assurance levels.

Many organizations and groups are integrated into the forensic DNA community. These groups include the International Society for Forensic Genetics (ISFG), the European Network of Forensic Science Institutes (ENFSI), the FBI's Scientific Working Group on DNA Analysis Methods (SWGDAM), and the Biology/DNA Scientific Area Committee of the Organization of Scientific Area Committees for Forensic Science (OSAC) in the United States. These groups are composed of practitioners, academics, and other parties who provide guidance, recommendations, and essential documentation to forensic laboratories. Laws governing DNA laboratories vary internationally, but some legislative controls bind most labs.

Accrediting bodies also audit laboratories for compliance with documentary standards, such as ISO 17025 from the International Organization for Standardization (ISO). Accreditation addresses many aspects of laboratory operations, from sample storage through analytical procedures, QA/QC,

Level	QA/QC Measure
Community	Quality assurance standards
Laboratory	Accreditation and audits
Analyst	Proficiency tests, certification, and continuing education
Method and instrumentation	Validation studies and analysis of control samples
Laboratory protocol	Documented standard procedures
Datasets	Positive and negative controls
Individual sample	Size standards and allelic ladders
Interpretation	Case review by a second analyst; QC review

Table 4.2 QA/QC measures in DNA analysis

documentation and reporting, and analyst training and continuing education. Accreditation includes self-study documents, site visits, reviews, and, once a laboratory is accredited, audits and periodic reaccreditation. In some situations, analysts may be certified, indicating they have successfully passed written and practical tests and participate in continuing education. Certification applies to the analyst, while accreditation applies to the laboratory.

Other quality assurance practices related to analysts ensure they have the educational background needed, extensive training and apprenticeship work under senior analysts, continuing education and training, and evaluation using proficiency tests. Proficiency tests may be generated within the laboratory or obtained from external organizations. Although it is desirable to create blind proficiency tests in which the analyst does not know it is a test, this is impractical in most cases. A blinded evaluation must mimic casework so closely that the analyst does not realize it is a proficiency test; this is a challenging task.

Instrument performance must be documented using positive and negative control samples. *Positive controls* have a known composition and DNA type, while *negative controls* have no DNA. Negative controls demonstrate that the system is free of contamination or carry-over from previous samples. For example, an extraction blank would be a clean, sterile swab extracted and tested the same way as any sample swab is analyzed by the laboratory. A negative control is produced when distilled water replaces the DNA extract. If DNA appears in the instrumental output of these controls, it means that somewhere in the procedure, contamination has been introduced. The source of this contamination must be identified and eliminated before any additional extractions proceed. Potential sources could include residual amounts of contaminating DNA on the swab or the small plastic vials used in the extraction, or in the reagents. Datasets from cases include many negative and positive controls integrated into the case and data review to ensure that nothing went wrong with the process from start to finish. Allelic ladders and size calibrations (Figures 4.5 and 4.6) are analyzed as part of each dataset.

Laboratories maintain documentation that specifies in detail how DNA analysis is performed from start to finish. Any deviation from the prescribed method, called a *standard operating procedure (SOP)*, must be documented in case notes. The SOP is reviewed periodically and updated as needed, with copies of previous versions preserved and secured. The specific procedure and version used in a case are integrated into the case record.

Once all the data is compiled and interpreted, a final review or reviews by someone other than the analyst occurs. This may be another analyst in the laboratory or a QA/QC manager or both. The reviewer is responsible for many tasks, including ensuring that the SOP was followed, that clear and complete documentation has been provided, that interpretation of the instrument output and assignment of each sample's genotypes are correct, and that QC requirements have been met. Only after this review is completed can results be released.

The Phantom of Heilbronn

A case that illustrates the dangers of contamination and the need for extensive QA/QC ended up not being a case at all. A prolific female serial killer thought to have committed dozens of murders did not exist, but it took years to realize this. The strange events were eventually linked to DNA contamination described as the case of the Phantom of Heilbronn or the Woman Without a Face. From 1993 to 2009, the same female DNA profile was recovered from evidence collected in dozens of crimes, including murder, that occurred across a wide area of Europe including Germany, France, and Austria. Investigation over the years showed no plausible linkage of the crimes and no discernible patterns. It wasn't until 2009 that the mystery was solved. The DNA found on the swabs was contamination that occurred in the factory where the swabs were manufactured. Proper sterilization procedures had been followed, but skin cells still found their way onto the swabs as trace DNA, a topic we will explore in detail in coming chapters. An ISO standard issued in 2016, "Minimizing the risk of human DNA contamination in products used to collect, store and analyse biological material for forensic purposes – Requirements," set forth procedures to avoid such problems. However, as we will see, as DNA detection improves, the potential for contamination interfering with the analysis and interpretation increases.

Interpretation

The term *DNA profile* refers to the combination of types found at the STR sites tested. As was the situation with ABO and isoenzymes, the frequency of a type is calculated by combining the types from each location (the "profile frequency estimate"). These calculations rely on frequency databases of the alleles in different populations, such as African-American and White.

As an example, look at Figure 4.6 and the D3S1358 STR site with a 16,17 type. A US White population database may show the frequency of these two alleles as 0.238 for allele 16 (16 repeated sequences) and 0.211 for allele 17 (17 repeated sequences). In other words, ~24% of the US White population is expected to have the 16 allele, and about 21% to have the 17 allele. The frequency of the 16,17 genotype is calculated by

$$2 \times 0.238 \times 0.211 = 0.1004$$

Thus, about 10% of the US White population is predicted to have this genotype at this STR locus (D3S1358). You can see how the discrimination power of this one STR locus already exceeds some of the ABO blood groups and most of the isoenzyme systems noted in Chapter 2.

Why is there a factor of two in the above expression? Recall our discussion from Chapter 1 regarding the allele frequencies using the *p/q* notation (Table 1.2). In the present example, there is only one way for a child to be homozygous 16,16 or 17,17. However, there are two ways for the child to have a heterozygous type of 16,17. The mother could contribute a 16 and the father a 17, or the mother could contribute the 17 and the father the 16. Thus, anytime there is a heterozygous type, the combined frequencies are multiplied by two.

The population database selected impacts the frequency results. Table 4.3 shows the 16,17 genotypes across several databases. It is crucial to appreciate that there are many databases, and that when new samples are added to them, frequency values change slightly over time. This is not a problem, since calculated frequencies are estimates and should always be presented and interpreted accordingly. The most noticeable difference is between Asian populations and the others. In practice, combined databases may be used

Database	16 allele	17 allele	Combined	%
Belgium	0.238	0.240	0.114	11.4
Denmark	0.225	0.255	0.115	11.5
EU combined	0.250	0.210	0.105	10.5
Finland	0.250	0.207	0.103	10.3
France	0.267	0.183	0.097	9.7
Germany	0.246	0.228	0.112	11.2
Norway	0.262	0.213	0.112	11.2
Sweden	0.241	0.229	0.111	11.1
Switzerland	0.246	0.205	0.101	10.1
Thailand	0.342	0.238	0.163	16.3
Asian combined	0.342	0.238	0.163	16.3
UK Black African and Caribbean	0.336	0.209	0.141	14.1
UK Chinese	0.328	0.254	0.166	16.6
UK Indian	0.308	0.215	0.132	13.2
UK White	0.241	0.205	0.099	9.9
US African American	0.319	0.212	0.135	13.5
US Asian	0.330	0.201	0.133	13.3
US White	0.238	0.211	0.100	10.0
US combined	0.283	0.204	0.115	11.5
US Hispanic	0.298	0.184	0.110	11.0

Sources: https://strider.online/frequencies (Europe and Asia); www.gov.uk/government/statistics/dna-population-data-to-support-the-implementation-of-national-dna-database-dna-17-profiling (UK); https://strbase.nist.gov/NISTpop.htm (US).

Table 4.3   Allele frequencies for D3S1358

when the ethnic identification of the contributor is unknown. Alternatively, results may be calculated for several subpopulations and all values reported.

The power of DNA typing comes from combining STR genotypes. Individual frequencies are combined using the product rule to obtain the overall frequency. The result can be astounding. For example, Figure 4.7 shows the electropherogram from an STR typing analysis for a White male from the United States. The probability of this profile occurring in a person drawn at random from this population is about one in nine quintillion. Table 4.4 illustrates the product rule calculations. The table order is arbitrary – what matters is the combined frequency.

The table calculations start with the same locus discussed above, D3S1358. The 16,17 type occurs in about 10% of the selected population or about one in ten individuals in the US White male population. The frequencies of the amelogenin site are not included in the calculation since the male population was specified. The rest of the table lists the STRs from Figure 4.7 in order from left to right, top to bottom. With the addition of the second STR, the cumulative frequency of the two geno-types is about 1 in 200, and it decreases rapidly from there to astounding numbers. In practice, numbers over a trillion may not be stated because they are so large; a quadrillion is a million billion, and a quintillion is a thousand quadrillion. By way of comparison, the total population of the world is around 8 billion people, which is 0.008 trillion.

Locus	Type	Frequency allele 1	Frequency allele 2	Combined frequency (one-in-x)	Cumulative frequency (one-in-x)
D3S1358	16,17	0.238	0.211	10	10
D8S1179	12,14	0.168	0.166	18	200
D21S11	28,30	0.159	0.283	11	2000
D7S820	9	0.168	0.168	36	71,000
CSF1PO	10	0.220	0.220	21	1.4 million
TH01	6	0.236	0.236	9	Millions
D13S317	11,14	0.326	0.043	36	
D16S539	11,13	0.314	0.026	60	Billions
D2S1338	22,23	0.035	0.105	137	
D19S433	12,14	0.071	0.362	20	Trillions
VWA	17,18	0.284	0.202	9	
TPOX	8	0.525	0.525	4	
D18S51	14,16	0.134	0.147	25	Quadrillions
D5S818	12,13	0.388	0.143	9	
FGA	21,22	0.179	0.205	14	Quintillions

Table 4.4   Combined frequencies for the DNA profile in Figure 4.5

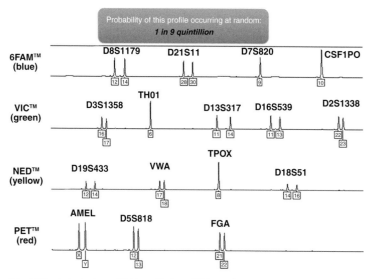

Figure 4.7 A single donor DNA profile for a White male.

The combined value of about one in nine quintillion is the random match probability for this profile using the specified allele frequencies. However, the random match probability has become one of the most misunderstood metrics in DNA typing because of the incredible numbers involved. In the current example, the probability of selecting a White male from the US population with this combination of STR types is much less than one in a trillion. It *does not* mean that there is only one male in the population with this type; there may be more. We cannot know the exact frequency; we can only estimate it. The allele frequencies used in this example provide this estimate based on the individuals typed in a specific set of population samples.

Unfortunately, because the values are so low, the random match probability has been misinterpreted as a probability of guilt, which it is *not*. Suppose the DNA profile from the example is found on a murder weapon and a POI with the same type is identified. When a random match probability is so low, it is tempting to think there is no other rational explanation for this finding other than guilt. It is crucial to realize that finding a given DNA profile does not tell us anything about how the DNA got there, when it got there, or how it got there. In other words, there is no direct link between a low random match probability and guilt.

Sexual Assault Cases and Mixtures

Sexual assault cases and vaginal swab evidence involve mixtures of two or more contributors. Differential extractions can usually isolate male and female fractions, but these are still challenging samples to analyze and interpret. The male and female fractions rarely contain comparable DNA amounts. If the man has had a vasectomy, no sperm cells are deposited although semen is present. The lack of sperm does not prevent DNA typing since other cells are present in a sample from a vasectomized male, but it does complicate sample preparation and interpretation, because differential extraction will not work. Often, traces of semen are found on the victim's clothing, bedding, or other physical evidence from the crime scene. Further complicating matters is the time that may have elapsed from collection to testing. A 2018 case from Japan is an example.

A high school student was sexually assaulted. She showered after the attack and reported to a hospital, where she was examined and evidence collected. A DNA profile was developed from an external swab. An investigation led to charges and a trial which started about a year after the crime. The defense was able to raise concerns regarding the identification. To address these issues, analysts obtained the skirt the victim wore when she was attacked. At the time of the crime, a semen stain had been noted but was not tested because positive results had already been obtained from the swab. The skirt had been returned to the victim, who had it dry cleaned. The stain was no longer visible, but analysts referenced case photos to locate the area where it had been. None of the screening tests for semen worked, but remarkably, sperm cells were found in the stained area. Analysts successfully generated a nearly complete STR profile from the skirt. Thus, even after a year had elapsed since the assault, DNA evidence was recovered from the skirt.

Chapter Summary

The current method of DNA profiling is based on short tandem repeats (STRs). These shorter repeating units are less prone to degradation than the sequences used in RFLP discussed in the last chapter. The ability to amplify STRs using PCR makes it possible to detect lower amounts of DNA as well. The ability to separate amplified DNA using capillary electrophoresis rather than the old slab gel method meant that DNA profiling could be automated, which allowed laboratories to dramatically increase the number of samples that could be analyzed in much less time than when using RFLP techniques. Finally, the ability to generate random match probabilities across many STR loci using the product rule provided a means to state results quantitatively with probabilities calculated from frequency databases. In the next chapter we will learn more about DNA data analysis and how simple DNA mixtures such as those seen in sexual assault cases are interpreted.

5 DNA Analysis and Interpretation: Single-Source Samples and Simple Mixtures

The last chapter discussed how peaks in an instrument output are converted into a DNA profile and how the random match probability is calculated using the product rule. Now we delve into how these profiles are analyzed and interpreted. Once a DNA profile has been developed from crime-scene evidence, it is compared to the profile(s) from known reference samples. These include elimination samples and samples from a person or persons of interest. If these comparisons do not provide helpful information, the profile can be submitted to a DNA database to search for investigative leads. Our focus is on DNA samples from a single person or simple mixtures such as a well-separated sample from a sexual assault case. Complex and low-level mixtures are much more challenging. We tackle those in the next chapter using the foundation we will build in this one.

Data Analysis and Interpretation

Several illustrations from the last chapter (e.g., Figures 4.5, 4.6, and 4.7) showed electropherograms generated for STR profiling. This section will discuss how peak-based data generates a profile and what steps are needed to ensure the data is usable and reliable. As will be discussed in this chapter, the electropherogram may be missing peaks (referred to as *allele drop-out*), have extra peaks (*allele drop-in* or *stutter artifacts*), or have misplaced peaks (*off-ladder alleles*) that must be considered and addressed. Instrument software converts the raw instrument output to interpretable patterns, but analysts carefully review the results and step in when issues arise.

What is a Peak?

Looking at Figure 4.7, you can see the peaks identified with numbers in boxes below them. This number is the allele designation, which is determined by comparing the base pair size (horizontal position) of the allele to the allelic ladder sizes (see Figure 4.6). The height (vertical position) of the peaks at each position reflects the amount of DNA present; peak height is essential for determining which electropherogram features represent alleles. You can see other small features in Figure 4.7 that could be peaks. For example, look at data associated with the first STR locus (D8S1179), upper left in the blue dye channel. You may see a blip between the 12 and 14 peaks and a blip to the left of allele 12. Next, study CSF1PO. There is another baseline blip to the left of the 10 peak. How can we be sure that this person is homozygous 10 at this location rather than having a 10 allele and another allele that did not produce a detectable peak? Laboratories utilize several techniques to sort genuine allele peaks from artifacts.

Any measurement system has an inherent level of background noise. This background appears in the output even if DNA is absent. Background noise is usually much smaller than sample signals. Laboratories characterize the background noise and generate signal-to-noise ratios through extensive method validation studies. The validation procedure incorporates experiments that prove the method is trustworthy and reliable and under what conditions. By knowing expected background noise levels, thresholds are established to differentiate authentic sample signals from the background.

Typically, two thresholds are set – a lower *analytical threshold* and a higher *interpretation threshold*, as shown in Figure 5.1. The baseline corresponds to the flat bottom line of an electropherogram, such as in Figure 4.5. If you zoom in to the baseline, you can see the small random fluctuations – this is the background noise. Any peak with a height that is too close to the expected noise levels is deemed unreliable. The analytical threshold is set based on the largest anticipated noise levels. Peaks below the analytical threshold are too close to the noise level to be trustworthy and are unreliable. Peaks above this level but below the interpretation threshold are typically used only for exclusion. This means that the peak can only be used to eliminate a POI as the source; it is not used to include that person. The reasoning for this is related to

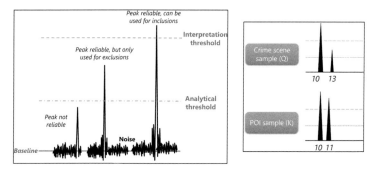

Figure 5.1   Left: Thresholds used to distinguish normal background noise from authentic allele peaks. Right: An idealized example of a crime-scene sample in which the peak of allele 13 falls above the analytical threshold but below the interpretation threshold. This peak can be used to exclude the person of interest (POI) based on this allele, but not to include others with a 10,13 genotype.

the potential for missing peaks associated with low levels of DNA. This phenomenon is call allele drop-out, defined under *Missing Peaks*, below. Peaks above the interpretation threshold are deemed reliable and used as part of inclusions. Note that the interpretation threshold (middle of Figure 5.1) is also called the *stochastic threshold*; we will discuss stochastic effects in Chapter 6.

Binary Thresholds

Thresholds such as those illustrated in Figure 5.1 are binary in the sense that we make a yes/no decision about interpreting a peak. This is a reasonable approach for single-source samples. Assume we have a case sample (Q) and a reference sample from a person of interest (K), as illustrated at the right in Figure 5.1. What if one peak from one locus from the crime-scene DNA profile falls between the analytical and interpretation thresholds, as does the peak for allele 13? We can use this peak (the 13 allele) to exclude the person of interest (K) because this profile does not have the 13 allele found in the crime-scene sample Q. This reasoning applies only to single-source samples and simple mixtures. Complex mixtures require much more extensive analysis, and methods that go beyond binary decisions thresholds.

Missing Peaks

Occasionally an authentic allele peak does not appear in the electropherogram. This situation is referred to as *allele drop-out*, and it occurs when an allele is not copied during the PCR amplification process. When allele drop-out occurs, the DNA profile from a heterozygous individual with two distinct alleles appears to be homozygous at this tested region. For example, if a person has a genotype of 8,14 at a locus but only a single 14 allele peak appears in the electropherogram, this means that allele 8 has been dropped. Allele drop-out results in a false homozygote designation at that STR locus, that is, a person who is heterozygous is assigned a homozygous profile.

Allele drop-out occurs either because a PCR primer cannot bind to the DNA template sequence correctly (due to a primer binding site mutation) or because there is an insufficient DNA amount available to be accurately copied (due to random effects). We discuss these effects in detail in Chapter 6. Recall from Chapters 3 and 4 that primers are designed to bind with the base pair sequence on either side of the STR locus of interest (Figures 3.1 and 4.4). If one of those bases mutates or differs in the tested sample, the primer cannot successfully bind to the DNA strand. If the primer doesn't bind, the sequence is not recognized and copied, and PCR amplification fails.

Low levels of DNA in a sample can also lead to amplification failure. Missing peaks are inherently more challenging to deal with than extra peaks, since it is difficult to know when a peak is missing in the first place. There are technical solutions in most cases, but the key is knowing that drop-out has occurred. The probability of an allele drop-out can be estimated using peak heights and complex algorithms based on statistical modeling. Probabilistic genotyping software (PGS) systems, discussed in Chapter 9, employ such algorithms.

Extra Peaks

Situations can arise in which there are extra peaks, misshapen peaks, misplaced peaks, or other problems with the instrument output. Software helps to identify such artifacts, which the analyst must address to ensure that an artifact is not identified as an allele or vice versa. Artifacts arise from many sources, including problems with the instrument, biological issues associated with PCR amplification, or some unique aspect of the person's DNA. Minor baseline blips such as those we discussed earlier in the chapter can come from dye residuals (a dye blob) or tiny air bubbles in the capillary tubes, for example. Voltage spikes in power lines can generate sharp signals in the electropherogram. Samples might contain trace-level contaminants that fluoresce, which can create peaks in the electropherogram. Most issues such as these are easily spotted and addressed by experienced analysts and noted in the case file.

A common artifact of biological origin is called *stutter*. *Stutter products* originate during the PCR amplification process when the polymerase enzyme copying DNA slips along the strand. This slippage results in a copy of the STR allele missing one of its repeat units. Stutter peaks are smaller in height than the genuine allele peak and differ in size based on the repeated base pair sequence. For example, if the repeating unit is the base sequence GATA, the stutter product will be displaced by four base pair units to the left of the allele peak in the electropherogram. Slippage is usually minor, resulting in stutter product peaks no taller than 15% of the genuine allele peak. When the stutter peak is more significant (as can happen when amplifying low quantities of DNA), it can be challenging to interpret the resulting pattern. STR markers with smaller repeating units of two or three base pairs are more prone to stutter peaks than STRs containing four base pairs or larger repeating units.

Another source of extra peaks may be off-ladder alleles. The electropherogram shown in Figure 4.5 is an allelic ladder for commonly typed STR loci. The "rungs" on these ladders are made by combining samples from multiple people to capture the range of expected alleles for each STR locus. However, rare or previously unknown alleles can appear in the electropherogram. Analysts flag the extra peak as a potential allele rather than an artifact such as a stutter peak. For example, look at Figure 4.6 and the allelic ladder for locus D3S1358. Suppose a sample produces a peak between the allelic ladder peaks labeled 16 and 17. This peak would represent an *off-ladder allele* that is larger than the 16 repeat unit allele and smaller than the 17 repeat unit allele. Another term used for such an allele, which does not align with the full repeat rungs on the allelic ladder, is a *microvariant*. If an STR marker has repeat units composed of four nucleotides (e.g., GATA), then off-ladder alleles could be labeled as .1, .2, or .3. The number to the right of the point is not a decimal fraction, but indicates the number of bases. For example, if the off-ladder allele was labeled as 16.2, the ".2" portion indicates that the detected peak measures two nucleotides longer than an allele 16. There would not be a 16.4, because in a tetranucleotide such as GATA, four bases longer would be the 17 repeat allele.

What Marker Does the Peak Represent?

At this point in the analysis of an electropherogram, we have established which peaks to interpret. Artifact peaks such as stutter and noise have been eliminated, and potential drop-outs flagged. The next step is to associate peaks with specific loci and alleles. Size standards are employed as part of this process. Figure 4.5 shows the common alleles for several STR loci. The bottom frame is an *internal size standard*. This sample contains DNA fragments of known sizes starting from 100 base pairs (bp) at the left through 350 bp on the right. The internal size standard converts measured peaks from migration time through the capillary (measured in seconds) to nominal bp size. The size standard and the dye color allow the data analysis program to assign STR locus identifiers to specific peak groups. For example, look at the four loci shown at the right side of Figure 4.5 (CSF1PO, D2S1338, D18S51, and FGA). All four amplification products are between 300 bp and 350 bp long, but each is tagged with a different dye color. Thus, the data system easily distinguishes between them.

The allelic ladder (Figure 4.5), analyzed with the same internal size standard, contains alleles of known length, and allows the analyst to assign peaks to specific alleles.

Q and K Comparison

Not all cases require a DNA database search. In situations where comparison profiles are available, Q and K profiles can be compared using the process described in Chapter 1. This comparison generally leads to reporting one of three conclusions – exclusion, inclusion, or inconclusive. Exclusion results if any allele differs between Q and K. If the Q and K profiles match at all loci, this is an inclusion, and a random match probability is calculated. Inconclusive results may occur due to the lack of a known reference sample (K profile) or sample degradation in the crime-scene sample (Q profile). To look for other potential persons of interest if there is an exclusion to the submitted POI, the crime-scene profile can be subjected to a database search.

DNA Databases

National DNA databases used for law enforcement purposes exist in many countries and are populated with DNA profiles from individuals required by law to provide samples. For example, convicted sex offenders must submit samples for DNA analysis, and these results are incorporated into law enforcement databases. These databases are separate from consumer genealogy databases. We talk much more about these sites and forensic genetic genealogy in Chapter 9.

The first national DNA database was established in the United Kingdom in 1995, based initially on six STR loci and then expanded to 10 in 1999. In 2011, the European Union began requiring a standard set of 12 STR markers to enable country-to-country comparisons when desired. While each country maintains its own national DNA database, crime-scene profiles may be searched against another country's set of offenders if DNA profiles contain a consistent set of STR markers.

The European Network of Forensic Science Institutes (ENFSI) produces surveys of European DNA databases across 45 countries, including Russia and Turkey. As of 2017, the cumulative size of various national DNA databases in Europe was over 11 million. Similarly, Canada implemented the National DNA Database of Canada in 2000, and Australia created a national database by merging smaller ones into the National Criminal Identification Database in 2001. A 2019 survey conducted by Interpol revealed that 70 of 194 member countries had DNA databases. Interpol established a database gateway in 2003 that facilitates profile exchanges and searches.

In the United States, the FBI selected 13 core STRs in 1997 after creating a national DNA database under a 1994 Congressional mandate. The federal–state–local nature of the US government and forensic laboratory structure led to the gradual adoption of the national clearinghouse model with the National DNA Index System (NDIS), which upon its launch in 1998 had seven participating states.

Today all 50 states, the District of Columbia, the federal government (including the FBI Laboratory), the US Army Criminal Investigation Laboratory, and Puerto Rico participate in NDIS using the Combined DNA Index System (CODIS) software to compare case-to-case or case-to-offender/arrestee profiles. It is crucial to appreciate that the US national DNA database is a network of linked systems, not one large centralized system. The primary nodes are the Local DNA Index Systems (LDIS) linked to State DNA Index Systems (SDIS), which in turn link to the national NDIS. Each state has different laws governing the types of crimes requiring a DNA sample to be collected. In some states, those arrested for violent crimes must submit samples.

Profiles in the NDIS are grouped into categories that are called indices. The two most extensive indices are the convicted offender index and the forensic index. The forensic index holds profiles that have been generated from crime scenes or other evidence. Most of these profiles are not associated with POIs; however, some are from older solved cases. These profiles may assist in linking crimes. Figure 5.2 shows the searching process for NDIS.

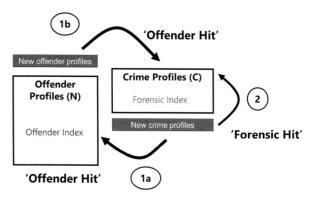

Figure 5.2   Searching modes in the US National DNA Index System (NDIS). Each index is a category of profiles. New offender profiles are searched against the forensic (crime-scene) index and new crime-scene profiles are searched against the offender and arrestee index.

When a laboratory submits a profile from a crime scene or other evidence, it is searched against the offender index (1a) and the arrestee index (not shown) to see if it links to any individual. New offender profiles are searched against the forensic index (1b) to see if they link to any crimes. Profiles from crime scenes or evidence are also searched against the forensic index (2) to see if it provides a link between crimes. An offender "hit" results if the submitted crime-scene profile matches an offender or arrestee at all loci. A forensic hit results if the submitted profile matches another crime.

As the number of DNA profiles has increased over the years, additional core loci have been added (much like new area codes when the number of cell phones has grown). Since 2017, the FBI has required 20 STR loci and the amelogenin site in profiles submitted to NDIS. As of October 2021, NDIS contains over 14.8 million offender profiles, 4.5 million arrestee profiles, and 1.1 million forensic (crime-scene) profiles. These 20 million profiles have assisted in more than 574,000 investigations.

The value of DNA databases is beyond dispute; populating them is not. Options range from a universal database in which all citizens would have profiles stored to minimalist databases where only convicted criminals' profiles reside. Questions range from which profiles to keep to how long to retain these profiles in the database. For example, several US states require that anyone arrested for a felony must submit a DNA sample. Should those profiles be deleted if the arrestee is cleared? These are the types of questions that arise.

There are several misconceptions about DNA databases. As an example, the CODIS system in the US stores STR profiles but no names. Only the STR profile is stored in a DNA database, along with a reference number. Names are stored in separate law enforcement databases. In the NDIS system, the offender and arrestee indices, which contain known individuals' profiles, are kept separate from forensic (crime-scene) profiles. Should a profile developed from a crime scene in a case match a profile in the offender or arrestee indices, the next step is to contact the agency that submitted the stored reference profile to obtain additional information about the person whose profile resulted in a hit. Additionally, databases associated with law enforcement in the EU, UK, and US do not contain any genetic information beyond the STR profile. As noted in the last chapter, STRs are in non-coding regions of DNA, do not encode for proteins, and thus do not control genetic traits such as eye color or blood type. STR types are highly variable inherited characteristics, but STR profiles reveal nothing regarding a person's health, susceptibility to diseases, or physical appearance.

Search Results vs. Conclusions

A database search produces one of three outcomes. First, a hit, in which the submitted profile matches a profile in the database at all loci; second, a partial match, in which some types are shared; and third, no matching genotypes at any of the loci typed and submitted. Search results are different from a conclusion. If the submitted profile matches a profile in the database at every locus, this is very strong evidence that the profile and the submitted sample came from the same individual. However, it does not prove it. The random match probability (Chapter 4) provides the necessary context to interpret any potential matches with a single-source crime-scene profile. There is always a non-zero probability that more people may have the same profile. It may be a vanishingly small probability, but it is never zero. Additionally, finding someone's DNA in a sample tells us nothing about how, when, or why it came to be there. The case of the Phantom of Heilbronn proved this point. We will see similar examples in the coming chapters.

Suppose a submitted DNA profile matches a profile in a database at all but one STR loci. This partial match proves that the submitted profile and the database profile could *not* have come from the same person (an exclusion). However, this does not mean that the results are useless. Similar STR profiles can be evidence of a familial relationship, such as between brothers or sisters. We discuss familial DNA in Chapter 9.

If a search produces no hits, the only unassailable conclusion is that there are no matching profiles in the database. Although DNA databases have grown significantly, include millions of DNA profiles, and grow larger every day, these numbers represent a small subset of the world population. The forensic index with crime-scene samples is regularly searched against new offender or arrestee profiles. For example, a forensic profile that produced no matches in 2004 might yield a hit in 2021 after additional reference profiles have been added.

Finally, a database hit might not lead to identification but rather link offenses. If a profile from a sexual assault is submitted to a national database and produces no hits, it is still in the database. If a year later, another profile from a different case in a different location is submitted to the database and matches the first one, this can establish that a common perpetrator was involved. However, if the perpetrator's profile is not in the offender or arrestee indices, the crime-to-crime match does not lead directly to an identification.

Simple Mixtures

So far, we have focused on single-source samples and simple mixtures such as found in single-assailant sexual assaults. The next chapter tackles complex mixtures, which present significant challenges in analysis and interpretation. We will focus here on two-person mixture combinations, which occur in many sexual assault cases. We will assume that these mixtures contain relatively large amounts of DNA, and that the contributions from both people are about equal. This treatment will provide the necessary background to tackle the complex mixtures to come.

Is it a Mixture?

The first issue in mixture interpretation is recognizing that a sample is a mixture. This is not always an easy task. Figure 5.3 shows one locus D18S51) isolated from a DNA profile. This comes from a known two-person mixture, but even if we didn't know this, we can quickly deduce it from peak pattern. Since there are three allele peaks at this locus, there is more than one contributor.

The enlarged D18S51 region in Figure 5.3 illustrates several points made earlier in this chapter. You can see the noisy baseline and two blips, one to the left of the 13 peak and one to the left of the 17 peak. These are likely stutter products. How do we know this? The STR repeat unit for D18S51 is AAGA, meaning stutter peaks will be found four base pairs to the left of the allele peak. The expanded region shows this offset on the length scale at the top of the frame. Notice how the 14 allele peak is not much taller than the stutter peaks, which shows that differentiating artifacts from genuine allele peaks can be difficult.

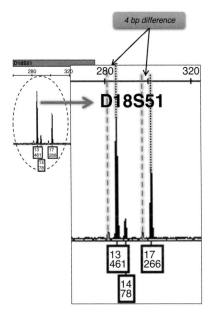

Figure 5.3 A portion of an electropherogram from a mixture. The box at right is an enlargement of the D18S51 locus. The vertical dotted lines are aligned with small peaks to the left of the larger peak. The offset in both cases is 4 bp units, which suggests a stutter peak.

Mixture Ratios and Peak Patterns

The combination ratio can be estimated from the electropherogram and the relative peak heights, although even a simple mixture presents complications. As the ratio becomes more unbalanced, say 1:5, it becomes challenging to differentiate artifact peaks such as stutter peaks from genuine allele peaks associated with the minor contributor. Once the minor contributor portion falls below about 20%, it is exceedingly tricky to develop a complete DNA profile for the minor contributor manually.

Figure 5.4 shows how peak patterns are employed to estimate the mixture contributor ratio in the electropherogram using a different locus, from the mixture, D2S1338. The presence of four peaks indicates that the contributors are heterozygous, which means that in a 50:50 mixture, we expect the peaks to be about the same height. Instead, we have two pairs of similar peaks, indicating that one person is type 19,24 and the other is 17,25. This is the correct identification for this mixture. The major fraction is calculated by comparing the peak sizes from the major contributor to the minor contributor. The process is shown in Figure 5.4. The number below the allele label is a measure of the peak height, and using this data, the major fraction is calculated and expressed as a percentage. This mixture is about 70:30, which means that approximately 70% of the sample comes from one person (the major contributor) and the remaining 30% comes from another person (the minor contributor).

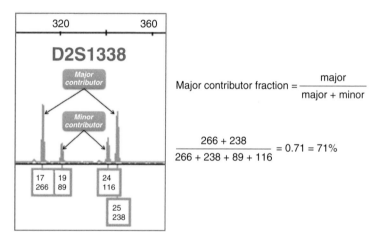

$$\text{Major contributor fraction} = \frac{\text{major}}{\text{major} + \text{minor}}$$

$$\frac{266 + 238}{266 + 238 + 89 + 116} = 0.71 = 71\%$$

Figure 5.4   At left is an enlarged view of the D2S1338 locus obtained from the two-person mixture. The fraction of the sample that has come from the major contributor is calculated as a percentage of the major contributor peak heights divided by the combined height of all peaks.

Once the mixture ratio is established, the information is used for interpreting peak patterns. Figure 5.5 illustrates how this is accomplished for D18S51. The genotypes for this known mixture are represented as triangles with heights approximating the 70:30 ratio; note that the 13 allele peak for the major contributor (70% of the mixture) is taller than that of the minor contributor (30%). Stacking the two triangles representing the 13 allele result in the tallest

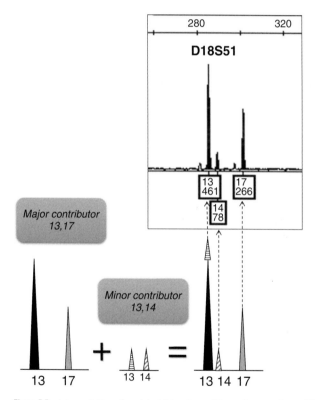

Figure 5.5   Interpretation of peak height patterns. The major contributor (13,17) is about 70% of the mixture and the peaks are shown in solid colors. The peaks from the minor contributor (13,14) are shown with dashed shapes. When combined (lower left), these contributions produce the electropherogram peak pattern shown in the upper right.

peak of the three. Thus, the diagrammed pattern is comparable to the actual peaks. We could not have deduced this pattern without knowing the approximate ratio of major to minor contributors.

These examples are from a simple mixture. Keep in mind that an analyst only observes the peak patterns. We used a known mixture sample to illustrate interpretation, providing an advantage that analysts do not have. Imagine how complex the interpretation becomes with mixtures of three or more contributors at different concentration ratios. The keys to understanding mixture data include the number of contributors, the relative contribution of each, and the peak height data. Data from all loci must be evaluated and any interpretation must be internally consistent. As a result, sophisticated software packages have been developed for mixture interpretation, as we discuss in Chapter 9. We won't go into any more detail here, but it is essential to appreciate how complicated mixtures can be and what challenges arise.

Deconvolution

Deconvolution is the process of separating a mixture profile into major and minor contributors. It may not be successful for all loci, but random match probabilities can be calculated when it is. With sexual assault cases, analysts typically have access to a reference sample from the victim. The data can be subtracted from the mixed profile to isolate the perpetrator's contribution. In addition, differential extraction methods (described in Chapter 4) are helpful for physically separating male and female fractions before amplification and profiling. Other types of evidence, such as mixed bloodstains, are not easily resolved into individual contributors unless reference samples are available.

Likelihood Ratios

We saw in the previous chapter how a random match probability is calculated and the astounding values it can reach, such as one in quadrillions. How this value is expressed is critically important for making the best and proper use of the information. Even one in a quadrillion is not definitive identification. It is powerful evidence for sure, but words such as "powerful" are inherently vague. In addition, the methods we have discussed up to now are reasonable for single-source and simple mixture samples, but with complex mixtures or low-level DNA samples, the random match probability falls short because this approach cannot account for the possibility of allele drop-out. An alternative metric is the *likelihood ratio* (*LR*). It can be applied to single-source samples and simple mixtures, but it is particularly useful for the more complex sample types we tackle in the next chapter. We do not need to detail how LRs are derived, calculated, or used, but a quick overview is beneficial.

The LR expresses the ratio of probabilities for observing the evidence profile given two specific scenarios called propositions or hypotheses. An LR of 1.0 means that the odds are even, and thus the test result is deemed uninformative. If the LR is greater than one, the results lend more support to the proposition or hypothesis in the numerator; if the LR is less than one, the results lend more support to the proposition or hypothesis in the denominator.

With forensic DNA typing, the LR is based on two competing hypotheses: the prosecution hypothesis (usually listed in the numerator; that the POI is in the mixture) and the defense hypothesis (usually listed in the denominator; that the POI is not in the mixture). Suppose a crime-scene sample generates a mixed profile from two individuals. One POI has been identified and has provided a reference sample. The LR is calculated as 1200. The result could be worded as "The results obtained from the crime-scene sample are 1200 times more likely if these results originated from the POI and an unknown, unrelated person than if the results originated from two unknown, unrelated people." Thus, with an LR of 1200, the evidence result provides moderate support if the hypothesis that the POI contributed to the sample is true.

The LR does not provide a definitive identification, just as the random match probability does not provide definitive identification. Using DNA data and specific propositions defined by the context of the case, the LR provides a statistical basis to express the results and frame the interpretation. One of the advantages of the LR method is that it affords a quantitative aspect to inclusion and exclusion decisions. In the example above, providing a likelihood ratio of 1200 is more informative than reporting that the POI could be a contributor (inclusion) without a numerical value reflecting the strength of the evidence. However, LRs alone should not be used to call results inconclusive. This decision arises from incomplete information or data.

Case Examples

This chapter introduces many essential topics that are best illustrated by examples and cases. The 2010 murder of a 13-year-old girl in Italy utilized a complex mixture analysis to identify the killer. It was discussed in a 2019 paper by Graversen and co-authors (see the reference list for the full citation). The girl left home one day to go exercise and disappeared. Her body was found months later. She had been sexually assaulted, stabbed, and left for dead, but the autopsy revealed that she hadn't died immediately; the cause of death was determined to be a combination of bleeding and hypothermia. Samples were collected from her underwear, and DNA analysis showed a male contributor to the mixture. The male profile did not match any database profiles, so investigators collected thousands of samples from males known to be in the area at the time of the murder, along with samples from the girl's relatives. No matches were found, but two profiles shared many alleles with the suspect, suggesting they could be related to the killer.

These tested profiles belonged to two brothers not involved in the crime. DNA collected from their mother showed that she could not be the mother of the killer. Their father died in 1999. Samples of his DNA were obtained from a stamp he had licked and from the exhumed body. These DNA profiles indicated that he was likely the father of the suspect. However, the two brothers were the man's only known children. Investigators suspected that the father had an illegitimate son unknown to the brothers or their mother.

The next step was their attempt to identify women who could have been the suspect's mother. This process involved a detailed study of the local population when the illegitimate child could have been born. Eventually, investigators located a woman who had lived in the area at the time who seemed a likely candidate. Analysis of her DNA showed that she could have been the mother of the perpetrator. The DNA evidence, along with extensive investigation, led them to her son, whose DNA profile matched that recovered from the girl's clothing. The son was convicted of the crime and sentenced to life in prison. The key to the case was the ability of DNA analysts to separate the crime-scene mixture, which involved extensive laboratory and statistical analysis.

A second initially perplexing case used LRs to aid investigators in another murder case in Tibet, published in 2019 by Xiao and co-authors. A man was found on a remote hillside bound with prayer flags. He had a massive head wound that appeared to have been caused by nearby bloodstained rocks. Conditions at the scene suggested that the man had been with sex workers before his death. He was not from the local area of Tibet. In addition to autopsy samples, DNA evidence was collected from the rocks, the prayer flags, and the penis. Investigators deduced that the man was likely Han Chinese, based on the use of the prayer flags. They felt it unlikely that a Tibetan would employ them sexually. This insight helped determine which populations to focus on during the database searches.

The penile swab mixture showed one female contributor. DNA samples were collected from women living in the area, leading to a POI, who provided a reference sample. An LR of more than four trillion was obtained from the evidence profile if this POI was considered a contributor to the mixture versus an unknown, unrelated woman. When questioned, the woman told investigators that she had had sex with the man the day before his death.

Complications arose with the analysis of mixtures from the prayer flags. Investigators reasoned that the perpetrator must have handled these during the commission of the murder, so this was vital evidence. A multi-contributor mixture of male and female was found on the flag. The female portion matched that of another sex worker from the DNA dragnet. However, the investigation proved she was not involved in the killing itself. Another LR was calculated, producing very strong support for the hypothesis that the combination found on the prayer flag was from the victim, the second woman, and an unidentified male. Police gathered blood samples of the men she had had sex with before the murder. One of those profiles matched that of the male contributor to the prayer flag mixture. Thus, this case involved complex mixtures, low-level DNA, and DNA transfer from a person not involved in the crime. This illustrates yet again that finding DNA on a piece of evidence tells us nothing about how and when it got there, regardless of how small the random match probability might be.

Chapter Summary

This chapter has described in more detail how data generated by STR analysis is interpreted. We learned how not all peaks are genuine allele peaks and what issues can arise with low-level DNA samples such as allele drop-out and peaks with heights below the analytical threshold. We introduced mixtures and how they can be revealed by the data, and saw how likelihood ratios are employed to assist in data interpretation. Finally, we examined two cases that involved many of the complexities of mixture cases. One reason such cases are common is that STR methods have become very sensitive and can detect very small amounts of DNA. The next chapter explores how this has been both a blessing and a challenge for DNA profile evidence.

6 The Curse of Sensitivity

The last chapter outlined the basic concepts of mixture analysis. Now we move on to the much more challenging situations arising from low-level DNA samples and complex mixtures. These topics go together. Early DNA methods such as RFLP and initial PCR methods were less sensitive (which means they were unable to detect very small quantities of DNA) than today's techniques. As a result, DNA present in tiny quantities was not seen. Now the technologies afford much better detection, which is a mixed blessing. Rather than simply detecting the DNA from the major contributor(s), now trace levels of DNA can be recovered and typed, and not all of it is pertinent to the crime under investigation. Very small amounts of DNA, much less than in typical samples, are referred to as *low copy number* (*LCN*) DNA.

Low copy number and complex mixtures are related. Think about a door handle on a busy office building. Dozens or even hundreds of people may touch the handle every day, and many will leave cellular residues and DNA. You would not expect any one person to be the dominant contributor to the mixture of DNA on the door handle. If the handle is swabbed and extracted, a complex mixture of many contributors, each accounting for a minute fraction of the total DNA present, would result. This type of situation frequently occurs in forensic analysis and happens now more so than in the past, given the rapid improvement in DNA sampling, preparation, and analysis technologies. As a result, small residuals of DNA are now detectable. This is why improvements in sensitivity are both a blessing and a curse – the more sensitive the method, the lower the detection limits and the more DNA detected.

Low Copy Number

Table 4.1 in Chapter 4 showed typical amounts of DNA in different types of evidence. Low copy number situations arise when less than approximately 100 pg (0.1 ng) of DNA is available. This amount corresponds to about 15 cells. The term *low template DNA* (*LTDNA*) is also used. LCN protocols typically involve increasing the number of PCR cycles to increase the amount of DNA amplified. This practice is not a foolproof strategy. Think of it as turning up the volume to the highest level. Yes, the music is louder, but so is any background noise. In many cases, distinguishing the STR results from amplified artifacts becomes paramount.

Stochastic Effects

We mentioned stochastic effects in the last chapter. Recall that stochastic effects are random and unpredictable. Such effects occur during the PCR amplification process and are particularly important with LCN samples and mixtures. Specifically, these effects arise because of unequal copying of the two alleles at a given locus. Primers fail to recognize and bind with the target sites, which in turn may cause a failure to copy an allele present in the sample. Figure 6.1 shows how stochastic effects arise and their impact on the electropherogram.

With high numbers of alleles available, PCR usually proceeds as shown at left in the top frame of Figure 6.1. With low DNA amounts (as in the top right frame), the sampling process may miss some of the few strands available, resulting in uneven copying and an allele imbalance in the electropherogram. The extreme case is allele drop-out, in which one peak disappears. There is no predominant cause for drop-outs, just a range of random effects related to the LCN problem. If a sample contains DNA from multiple contributors, drop-out may not be obvious.

The lower four frames of Figure 6.1 show examples of other types of stochastic effects, all due to failures in the PCR stage. In each case, the true genotype is shown in the box. We expect the two allele peaks to be close to the same height in the absence of stochastic effects; note the severe peak imbalance shown in the bottom left frame. The next frame to the right shows drop-out of the 14 allele.

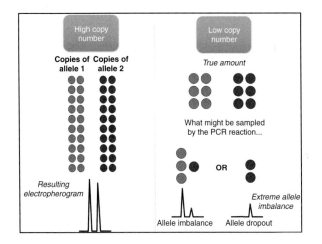

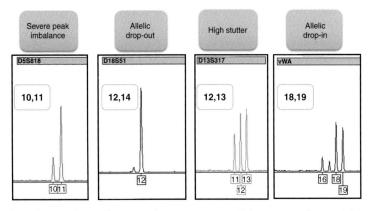

Figure 6.1   Top frame: Illustration of PCR copy issues with low copy numbers. At left is a sample with typical DNA amounts in which copy problems are minimal and the electropherogram peaks are normal. The right side illustrates uneven copying efficiency and the resulting allele imbalance. The lower four frames illustrate other stochastic effects, with the true genotype shown in the box.

The following two frames illustrate stochastic effects that generate extra peaks. High stutter is more common with low copy numbers and, in this example, has produced an additional peak that appears at the same place as the 11 allele. The location makes sense, since a stutter means that the enzyme used to unzip the DNA has slipped by one repeating unit on the DNA strand. Previously, we have seen small stutter peaks quickly flagged as artifacts, but LCN samples make this more difficult.

Allelic drop-in, shown in the far-right frame, is another effect, and the most common cause is elevated stutter. Here you can see what could be a stutter peak at the allele 17 position (unlabeled peak between 16 and 18) and an allele drop-in at the allele 16 peak position. Contaminants are another source of false peaks and allelic drop-ins. When abundant amounts of DNA are present, these spurious peaks are small enough to be easily identified as artifacts. When the relevant DNA is scarce, the size of allele drop-in peaks can approach those of genuine alleles.

Fortunately, there are steps and procedures analysts can use to counter some stochastic effects. Since drop-ins are often caused by trace contamination, they are usually not reproducible. Repeating the amplification process often eliminates the artifact peaks. For example, a laboratory may perform three amplifications of an LCN sample and only accept a peak as genuine if it appears in two of the three results. Of course, this technique only addresses random effects. If a contaminant originates from a crime-scene sample, it will be present in the initial extract. It will be amplified no matter how many different times the process is attempted.

Working with LCN

Improved detection capabilities impact forensic laboratories in many ways. Facilities must emphasize cleanliness and sterilization at all stages and places of evidence handling, from collection through storage and analysis. Those working in the laboratory are potential sources of trace DNA, and as such, their DNA profiles are kept on file to ensure that they are not DNA contributors to case samples. Other steps that can be taken include pre-concentration of the extract to small volumes, which increases the concentration of the DNA in the solution, and selective filtration to reduce the number of potential interferents in the sample extract before the amplification process. Finally, LCN situations mean that it may not be possible to extract enough DNA from the evidence to obtain sufficient DNA without consuming the entire sample. If that is the case, there would be no material available for any future testing of the evidence, a situation that is avoided when possible.

What is Touch DNA?

Touch DNA samples combine complex mixtures with LCN DNA. As the name implies, touch DNA is deposited when an object or material is handled or touched. In many cases, a DNA profile can be developed from a fingerprint because touching leaves behind skin cells and cellular debris. Trace quantities of DNA have been found and exploited from many types of samples, including fragments of improvised explosive devices, ammunition cartridges, skin surfaces on the necks of strangulation victims, paper, wiring and cables, firearms, clothing, and computers and peripheral devices. These are in addition to sources we discussed in Chapter 2. Just because DNA is present doesn't mean it has value in a criminal investigation, as was illustrated with the example of the door handle of an office building. Sorting relevant from irrelevant information in such situations is central to the analysis and interpretation of touch DNA.

Sources of Touch and Trace DNA

The ability to detect trace levels of DNA on so many surfaces means that it shares characteristics of trace evidence. Traditional types of trace evidence include hair, fiber, glass, and gunshot residue. Let's say you own a cat. Your clothing will have cat hair adhering to it. Some of these hairs transfer to surfaces such as the carpet and seats when you are in the car. If a friend rides in your car, their clothing may pick up cat hair that they can then transport to their own home. Even if they do not own a cat, cat hair from your cat will be on their clothes and in their house.

This situation arises through a series of trace evidence transfers. In this case, the initial deposit (your cat shedding a hair) is the primary transfer, while successive transfers are indirect transfers. When your clothing captures cat hair from furniture in your house, this is an example of a secondary transfer. Hairs drop off your clothing and onto the car seat due to tertiary transfer. Additional transfers occur when your friend sitting in the car picks up cat hair and when your cat's hair ends up deposited on surfaces in your friend's home.

Each successive transfer is smaller than the last. If your clothing gathers 100 hairs, perhaps 50 of those will end up in the car, 25 on your friend's clothing, and 10 in their home. The further the transfer is from the initial deposit, the less of the original material remains. The transfer process is known as Locard's exchange principle, paraphrased as "every contact leaves a trace." Edmond Locard (1877–1966) was a French criminologist who worked with trace evidence and published extensively in the field.

The same ideas regarding transfer (discussed in Chapter 2) apply to DNA, although DNA does not transfer quite as easily as cat hair. The original DNA source may be skin or biological fluids such as blood, saliva, or semen. Blood dripping from a wound onto a surface is an example of a primary or direct transfer of DNA. Touching a surface also results in a direct transfer because skin cells and DNA are deposited. If blood drips on a floor and someone walks through it, blood can be transferred to their shoes. Sequences of transfers are possible as well. A transfer is more likely with biological fluids when the fluid is wet than when it is dry.

Other factors can influence skin contact transfers. One example is that freshly washed hands will not transfer as much DNA as dirty, sweaty hands. DNA can be transferred in two directions, as is the case if two people shake hands. If you review the most recent few hours of your day, you can probably identify many instances where DNA transfers could have occurred. Thus, finding DNA on an object is not unusual; in fact, it would be unusual *not* to find it.

Locating Trace DNA

You might have identified one of the conundrums of trace DNA – how to find it. When evidence consists of a visible bloodstain, finding DNA is as easy as finding the stain. Low-level DNA, including that deposited from skin contact, does not involve biological fluids. In some cases, a fingerprint shows what area to search, but this is not the typical scenario by which trace levels of DNA end up on a surface. Specialized dyes and visualization methods have been developed recently to aid in locating potential DNA sources. Examples are shown in Figure 6.2.

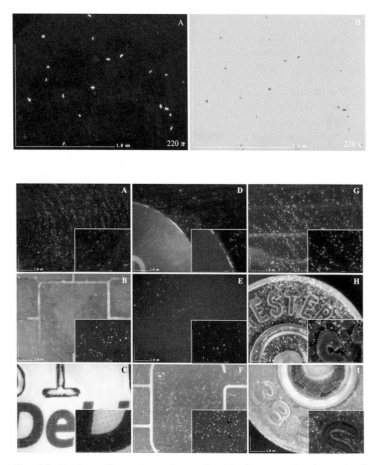

Figure 6.2   Top frame: Fingerprint on glass; on the right, the untreated surface with cells visible; on the left, the same area treated with a fluorescent dye. Bottom frame: Various surfaces treated with the dye. See text for description.

The top frame of the figure shows part of a fingerprint placed on a glass slide. In the left image it has been treated with a fluorescent dye sensitive to nucleic acids. Cells and cellular debris are evident in both treated and untreated images, and the dye differentiates cells from other debris. Where skin cells are present, DNA will also be there. The fingerprint is helpful as a control since pressing a finger onto a glass slide will transfer skin cells. The versatility of the dyeing method is illustrated in the lower frame, where the fluorescent dye was used on several surfaces:

A   Glass slide
B   Credit card metal chip
C   Credit card plastic
D   Home button area of a mobile phone
E   Mobile phone screen
F   SIM card
G   Plastic bag
H   Base of a nickel ammunition cartridge
I   Base of an aluminum ammunition cartridge

Fingerprint patterns are evident in images A and G in the lower frame of Figure 6.2 (glass slide and plastic bag, respectively). The dye used in this study does not react with DNA found in bacteria, which is valuable information that can eliminate unnecessary sampling. The dye guides the sampling of the locations where DNA is most likely to be found. This process is called *targeted* or *directed sampling*. This is important, because time and costly reagents are consumed with each sample tested. The dye does not interfere with subsequent DNA analysis. Recall that about 15 cells are the minimum needed for robust DNA analysis; as you can see in Figure 6.2, there is plenty of DNA on all of these surfaces for typing.

Relevant Transfer, Background, or Contamination?

The title of this chapter hints at the promise and cautions that apply to minute DNA quantities such as touch and transfer DNA. Finding DNA on evidence reveals nothing about how it got there, an issue that becomes paramount with LCN DNA evidence. Consider the case of Lukis Anderson, a homeless man who was arrested for a murder that took place while he was hospitalized, extremely drunk, and nearly unconscious. In late 2012, a wealthy Silicon Valley entrepreneur was murdered in his home. DNA evidence matched Anderson's profile, leading to his arrest. Hours before the killing, paramedics had responded to a call to treat Anderson, who was taken to the hospital and placed under constant medical supervision. Several hours after treating Anderson, the same paramedics responded to the murder scene. They used the same finger oxygen monitor for both men and had extensive physical contact with both. Thus, Anderson's DNA found its way to the crime scene via indirect transfer. Significant time, effort, and an investigation were needed to sort out the situation. Had Anderson not been hospitalized or not had an indisputable alibi, the outcome could have been much worse.

As we have seen, DNA is easily transferred and may end up a long way (in space and time) from where it originated. Suppose a minute bloodstain is found on a kitchen knife handle. Analysis of the stain produces a complex DNA mixture profile. This is not surprising – more than one person in the household may have used the knife for food preparation. The knife could have been stored in a frequently accessed drawer, providing more opportunities for transfer. The critical factor for separating relevant from irrelevant DNA is the time of deposition. A transfer that occurred during the commission of the crime is of evidentiary value; all other transfers are irrelevant. DNA deposited before the crime is called *background*, while DNA deposited afterward is called *contamination*.

The timeframe surrounding a crime can be divided into four intervals: before the crime; during the crime; after the crime but before the scene is secured; and after the scene is secured. DNA evidence created during the act is relevant, and there are techniques to isolate this critical period. Assume a bathroom counter is the surface of interest. Investigators postulate that the perpetrator of a murder was injured during the attack and deposited blood on the counter. The surface appears to have been recently wiped down. Counters are frequently touched, so background DNA traces will likely be present, contributed by household members and guests. Samples taken from areas on the counter and in the bathroom can determine the background DNA. At the same time, investigators can develop lists of persons who probably contributed to this background DNA. Reference samples are obtained and utilized during the investigation. Similarly, DNA deposited after the scene was secured is identified through elimination samples. Elimination samples are routinely collected from those present during the scene investigation and from anyone who handled the evidence during and after collection. Elimination samples from laboratory personnel are also available.

This combination of investigative procedures and DNA samples can help reveal what DNA was deposited before the criminal act and after the scene was secured. The combination of DNA sampling and investigation makes it easier to isolate the relevant DNA contributions that occurred during the crime or immediately after. Interpretation may not be simple, but without these steps, it might not be possible at all.

Shedders

We discussed secretors in Chapter 2; they excrete genetic markers such as their ABO blood type into body fluids. About 80% of us fall into this category. The term *shedder* has been used with touch DNA, although the concept is controversial. A good shedder tends to deposit more cells and cellular debris with touching than a poor shedder. A person's secretor status is often discussed as a binary descriptor – either you excrete blood group and isoenzyme proteins into your body fluids or you do not. However, shedder status is not as clear-cut, and no quantitative scales have been established. Shedder status seems to be more of a continuum, with most people falling into the middle range.

Dangers of Contamination

The ability to detect ever-smaller amounts of DNA increases the risks associated with contamination. A 2018 report from Switzerland noted that between 2011 and 2015, 709 instances of contamination of DNA evidence were reported, corresponding to about 1% of DNA profiles submitted for database searches. Of these, about 86% arose from police officers and 11% from laboratory workers. Contaminations involving police officers mainly occurred at the time of collection and were from the officer collecting the evidence. Follow-up interviews revealed that police did not always wear gloves and masks while collecting evidence. The study led to several recommendations to reduce contaminations in the field and the laboratory.

Gloves mitigate contamination risks but don't eliminate them. Wearing the same pair of gloves for several hours while handling multiple surfaces and objects is not much better than no gloves in terms of contamination risks. Gloves must be changed frequently and properly. Contamination risks in laboratories differ from the risks at crime scenes in that laboratories routinely handle DNA evidence and use controls and quality assurance samples. The cleanliness of surfaces, equipment, and storage facilities is critical and must be constantly maintained. Protective clothing beyond gloves is needed, and these items must be frequently and carefully changed.

Even the most rigorous practices and procedures cannot eliminate contamination. We have discussed how easily DNA transfers. Couple this fact with the sensitivity of DNA typing methods, and instances of contamination become unavoidable, even within the laboratory. One such case of laboratory contamination occurred in 2011 in the UK. In this situation, a single DNA profile was the only evidence against the accused. A woman was raped in Manchester, and a rape kit was collected. A swab was processed and yielded a DNA profile that matched a man who claimed he was in Plymouth at the time of the attack and had never been in Manchester. He was arrested and jailed. Eventually, mobile phone records showed that he had made calls from Plymouth during the time in question. He was cleared and released after five months in custody. An investigation revealed that his swab sample was contaminated by an unlikely series of events in the laboratory. The wrongly accused man was involved in an earlier, unrelated spitting incident, resulting in sample collection and submission to the laboratory. Incredibly, the same disposable plastic tray was used for both samples, resulting in the rape case swab contamination.

It's Not Just Whose DNA It Is, But How It Got There

A 2017 study evaluated the transfer and persistence of touch DNA on knife handles similar to the knife example discussed above. Study conditions were designed to mimic handling of regularly used kitchen knives. The knife handles (36 total) were thoroughly cleaned at the outset to remove any background DNA. Study participants provided DNA samples, as did their significant others. Each participant was assigned 12 knives, for which they were designated as the regular user. They handled the knife for two days as they would in typical cooking and food preparation. Indirect transfer was evaluated by having each person shake hands with other participants. Immediately after, each participant used their knife to stab a foam block over one minute repeatedly. Samples were collected before the handshakes and stabbings and at intervals from an hour to a week later.

DNA analysis showed that, as expected, profiles from the primary user were obtained from all the knives, although in different amounts. One person consistently left more DNA while another left smaller deposits that resulted in partial profiles. Results such as this led to the concept of good and poor shedders. Extraneous DNA was found in fewer than 5% of the cases in which only one participant handled the knife and there was no handshake. The exception was one participant; DNA from their significant other constituted about 25% of the DNA on the knife handles. After participants shook hands, DNA from the person whose hand was shaken made up about 10% of the recovered DNA. Study results were clear evidence of indirect DNA transfer.

The Amanda Knox Case

The infamous case of Amanda Knox involved DNA on a knife and illustrated the challenges and problems of touch and transfer DNA. The case emphasized that finding someone's DNA on a piece of evidence does not tell you how it got there. It also highlights the misconceptions that a match is the same as identification, and that identification corresponds to guilt. The saga began in 2007 with the murder of Meredith Kercher in Perugia, Italy. Meredith and Amanda shared an upper-floor apartment in a small house with two other young women. The floor below had an apartment shared by four young men. One of these men was Raffaele Sollecito, with whom Knox had a romantic relationship. She spent the night of November 2nd with him and returned to the apartment in the early morning. Upon entering the shared bathroom, she noted bloodstains in the sink. Kercher's door was locked. Knox went back downstairs and asked Sollecito for assistance. Sollecito entered the apartment and attempted to force the bedroom door open, without success. The two called police, who forced entry into the bedroom and found the body. Kercher had been stabbed multiple times. No potential murder weapons were found in the apartment.

During the investigation, police interviewed Kercher's boyfriend, Rudy Guede, who lived in an apartment in a different building. He initially denied any role in the killing but later claimed that he, Knox, and Sollecito had carried out the murder together. Police recovered a kitchen knife from Sollecito's apartment several days after the crime. The knife was eventually put forward as the murder weapon even though no blood was detected on it using screening tests.

The knife became critical evidence against Knox and Sollecito, and many of the lessons from the study we discussed above became critical. The prosecution alleged that the death resulted from a sex game gone wrong and that the knife was cleaned with bleach. This explained why the victim's DNA on the blade was found at such low levels. The accusations painted a salacious picture, and a media frenzy surrounded the investigation and legal proceedings. Knox, Sollecito, and Guede were initially convicted; in 2015, Sollecito and Knox were cleared.

DNA matching Knox was found on the knife handle. Low copy number DNA was found on the blade, but it produced an incomplete profile. The alleles that were identified matched the victim. Recall we noted that repeated amplification is one method to deal with stochastic effects. However, no replicate amplification was performed in this case. Subsequent analysis of the blade revealed starch, which is consistent with typical kitchen use such as slicing bread. It is hard to imagine a scenario in which vigorously cleaning the knife with bleach to remove all traces of blood would leave behind trace DNA and starch granules.

Many issues came to light with the knife, including improper collection techniques and a failure to conduct proper background testing. The low amounts of DNA found on the blade, coupled with the intermingling of the residents of the two apartments, supports transfer and contamination as a more likely explanation of the DNA results. Many other problems were highlighted in later reports, including poor handling of the knife from collection forward, many opportunities for cross-contamination by police and in the laboratory, and lack of background control samples. None of the other utensils in the cutlery drawer were tested, which would have shed light on the potential for indirect transfer.

As with the knife and Knox, only one piece of evidence could link Sollecito to the crime, which was Kercher's bra clasp. The bra was not collected until weeks after the murder. During this time, it had been handled and moved several times with no clear chain of control. Analysis of the clasp showed a mixture that matched Sollecito's profile at several loci. Sollecito touched the door while trying to gain entry the morning after the killing; subsequently, many police officers touched the door, entered the room, and handled items within. The doorknob was not tested, which could have provided valuable information regarding who had recently been in the room. The findings are more consistent with contamination and innocent transfer before or after the crime, rather than deposition during the crime.

Finally, DNA evidence was found that implicated Guede, including from the vaginal swab, blood on the bra strap (which had been cut off), and on a bloodstained sweatshirt and purse. None of the samples from the victim's room contained Knox's DNA. One theory advanced was that Knox selectively cleaned the room to remove her DNA but leave Guede's. This idea is as implausible as it sounds. Dr. Peter Gill, an internationally recognized expert in DNA, wrote in a 2016 report:

> Once DNA is transferred to an object, it may transfer to other locations on that object or to an entirely different object. The only way to ensure that one has not left DNA behind in a room is to clean it with sterile implements while wearing protective gear. It is not possible to clean a room to selectively remove all traces of one individual's DNA whilst leaving behind another individual's DNA. Moreover, in a murder in which three people allegedly participated in killing a victim who was fighting back, each of the perpetrators' DNA could be anywhere in the room. Saliva, which is rich in DNA, for example, could be sprayed on different surfaces. It would not be possible for one of the perpetrators to even know where his DNA as opposed to the DNA of others was deposited. The perpetrators' DNA could also be mixed together and it would not be

possible to somehow separate the DNA using household cleaning supplies.

The Knox case exemplifies the potential dark side of trace DNA and mixtures. Finding DNA somewhere tells us nothing about how it got there. Sorting that part of the story requires much more data, both from the laboratory and the investigation.

Tools and Techniques

Low levels of DNA can be collected in several ways. One is swabbing (dry and moist), as we noted in Chapter 4. Taping can be used with items such as clothing. A modified wet vacuum system called the M-Vac has been employed to recover trace DNA from porous surfaces such as wood and bricks, and, in one case, from a rock used to beat a woman to death. The device resembles a small carpet cleaner with a supply of sterile water directed through a brush handle. The pressurized water penetrates surfaces and is collected by vacuum into a filtered collection vessel. DNA in the solution is concentrated and extracted.

One of the first cases in which the M-Vac was used was a crime from 1995. The victim, Krystal Beslanowitch (age 17), was found alongside a river in Utah. She had been beaten to death. The murder weapon was a bloody rock found near the body. She lived and worked as a prostitute in Salt Lake City, about 45 miles from where her body was found. Her boyfriend and a taxi driver were persons of interest, but were eliminated during the initial investigation. The case went cold, but the rock was retained as evidence.

Eighteen years later, an M-Vac was used to collect samples from the rock, and a complete male profile was developed. It matched the profile of Joseph Simpson, who had been convicted of an earlier murder and had served time in Utah. At the time of the killing, he was employed as a shuttle driver at the airport. He drove a route that took him near the area where Beslanowitch worked. He left the state in 1997 and went to Florida.

Once the profile from the rock was matched to Simpson, detectives traveled to Florida and put Simpson under surveillance. They were able to recover a cigarette butt that Simpson threw away, and the DNA from this evidence matched that from the rock. Simpson was arrested and convicted in 2013.

Chapter Summary

The ability to detect low amounts of DNA (low copy number) inevitably leads to the detection of more mixtures. Touch and trace DNA that would not have been detected with methods used in the 1990s are now seen in many samples. DNA can be on a surface long before it becomes relevant in a criminal investigation by transfer or through contamination. Sorting out relevant from irrelevant DNA contributions has become a central challenge in DNA mixture interpretation. The Amanda Knox case illustrates the danger that arises from trace DNA and reiterates the point made previously that finding DNA on a piece of evidence does not tell us how it got there or when it got there; nor does it mean the person that it came from is guilty of a crime.

The last two chapters of the book will explore some new tools and techniques that are being used in mixture interpretation and LCN situations. First, we will take a slight detour to talk about DNA that is not inherited in the same way as STRs. This type of DNA can be useful in situations in which STR profiles are difficult to recover, such as in skeletal remains and missing persons cases.

7 From Mothers and Fathers

DNA From One Parent

So far, we have focused on DNA types in which one allele is from the father and one from the mother. However, three other sources of DNA come from only one parent, and all can be employed in forensic testing. One is mitochondrial DNA (from the mother in all her children), and the other two are STR sites on the Y chromosome (from the father in his sons) and STR sites on the X chromosome (from the mother in her sons). These DNA sources are *lineage markers*, since they can be traced back generations through our family trees. Lineage markers are valuable in missing person cases where DNA from the person of interest is not available. *Mitochondrial DNA (mtDNA)* has been used in historical cases, such as identifying soldiers killed in past conflicts. We will explore these and other examples in this chapter.

Lineage markers pass unchanged (except for mutations) from generation to generation and do not shuffle and recombine within each generation like the STRs we have discussed so far. This characteristic prevents the use of the product rule for calculating combined probabilities. Accordingly, lineage markers are less helpful in identifying specific individuals. In historical cases, this information may be sufficient to answer the pertinent question, but obtaining match probabilities of one in millions or billions is rarely possible.

Some new terms apply to lineage markers. STRs are described as genotypes, one allele from the mother and one from the father. Since lineage markers are from only one parent, they are referred to as *haplotypes*. A *haplogroup* is a set of haplotypes that are associated and thus tend to be inherited as a group. Haplogroups can be a valuable genetic tool in tracing human ancestry and tracking human migration patterns. Haplogroups can be characterized using *single nucleotide polymorphisms* (*SNPs*), which are discussed in the next chapter. Another term you will see is *autosomal DNA*, which refers to DNA found in the nucleus and on all chromosomes except the sex chromosomes (X and Y).

Patterns of Inheritance

Table 7.1 shows inheritance patterns of lineage markers compared to nuclear DNA STRs (Chapter 4). Genotypes of the autosomal STRs arise from the contribution of one allele from the mother and one from the father at any given locus. Thus, the first generation has 50% of their autosomal STR alleles from each parent. The amount of this contribution is halved with each successive generation. The inheritance patterns of the X and Y markers arise from the

Relationship	Nuclear DNA	Y chromosome	mtDNA	X chromosome
Mother ➜ son	50%	N/A	100%	100%
Mother ➜ daughter	50%	N/A	100%	50%
Father ➜ son	50%	100%	0%	0%
Father ➜ daughter	50%	0%	0%	100%
Paternal grandmother ➜ granddaughter	25%	N/A	0%	100%
Maternal grandmother ➜ granddaughter	25%	N/A	100%	25%
Paternal grandfather ➜ grandson	25%	100%	0%	0%

N/A (not applicable) indicates that the genetic marker is not found in the parent.

Table 7.1   Comparison of inheritance patterns for various genetic markers

passage of sex chromosomes from parents to children. Males are XY, and females are XX.

The mother passes one of her X chromosomes to all her children. If the child is a son (XY), the father's Y contribution is 100%. If the child is a daughter (XX), one X is from the mother, and one is from the father (50% from each). The mtDNA does not change (except for mutations) down the maternal line. Similarly, the Y markers do not change (except for mutations) in the male paternal lines (male children). We look at each type of lineage marker in the following sections.

What is Mitochondrial DNA?

So far, we have focused on DNA from the cell's nucleus, but as we saw in Chapter 1 (Figure 1.1), the mitochondria are another source of DNA. These structures are involved in energy production within the cell. Each cell has many mitochondria, meaning that each cell contains an abundance of mtDNA. This is one reason that mtDNA persists much longer in samples compared to nuclear DNA. mtDNA is double-stranded and circular. It is smaller than nuclear DNA and contains genes coding for proteins used in the mitochondria as part of energy production. Mitochondrial DNA differs from autosomal DNA because it is not found in the nucleus or on any chromosomes.

At conception, the sperm enters the egg, fertilizes it, and forms a *zygote*. When the zygote cell divides, the cytoplasm and other cell parts except the nucleus are consistent with the mother's egg cell. Thus, mitochondria and their mtDNA molecules are passed directly to all offspring independent of any male influence. Barring mutation, mtDNA passes unchanged from mothers to their children. Thus, any children in her line will have the same mtDNA.

As with nuclear DNA, there is a non-coding region that includes the areas targeted in mtDNA typing. This non-coding region of mtDNA (the control region) is divided into three *hypervariable* (*HV*) regions – HV1, HV2, and HV3. HV1 and HV 2 are commonly utilized in forensic testing. Hair, bone, and teeth are good sources of mtDNA, especially when examining decomposed or damaged remains. One of the advantages of mtDNA is that it can be found in naturally shed hair, whereas a root is needed to attempt typical STR profiling.

Early Cases

The first US criminal case involving mtDNA occurred in 1996. A four-year-old girl was raped and murdered in Tennessee, and a man named Paul Ware was discovered drunk and asleep next to her body. There was no semen found in her vagina or on her body, and none of her blood was found on Ware. The autopsy revealed one small red hair in the girl's throat, similar to several others located in a bed in the home. Reference samples for mtDNA analysis were collected from the victim (blood) and the suspect (saliva). The FBI conducted the analysis. The profiles from the hairs matched Ware and not the victim; this meant Ware could not be excluded as the source. Frequency estimates were obtained from a small database (742 people) the FBI had developed. Ware was convicted for rape and murder.

Another US case, the high-profile trial of Scott Peterson in 2003, also involved a single hair. Laci Peterson, pregnant at the time, disappeared in December 2002. The two badly decomposed bodies (mother and unborn child) were recovered in April the following year. Scott Peterson claimed to have gone fishing the day she disappeared. A search of the boat revealed one hair found on a pair of pliers, which was subjected to mtDNA analysis. The reference sample was hair from Laci's hairbrush. The profiles matched, and the results were admitted for trial. The judge specified that the results had to be properly stated; the mtDNA profiles were expected to be found in one person in 112. This is an essential point regarding mtDNA – it is not ideal for identification because the frequencies are relatively high compared to nuclear DNA and STR profiles. Just because it is DNA evidence does not mean it provides definitive identification. We will talk more about what the profiles look like and the statistics of mtDNA shortly, as they are significantly different from STRs. Peterson was convicted based on all the evidence, including the mtDNA.

Sequence Data

Mitochondrial DNA is characterized base by base (i.e., by its sequence), rather than by repeating units of bases. We will delve into sequencing in more detail in the next chapter. The base pair sequence of mtDNA was established in 1981

	Location and sequence within the HV region			
Sample	**16091–100**	**16101–110**	**16111–120**	**16121–130**
rCRS	ATTTCGTACA	TTACTGCCAG	CCACCATGAA	TATTGTACGG
Test sample	ATCTCGTACA	TTACTGCCAG	CCACCATGAA	TATTGTACAG

Table 7.2   Example haplotype assignment

by a team led by Dr. Frederick Sanger in Cambridge, England. The sequence consists of about 16,500 base pairs. Length can differ based on mutations that drop or add base pairs. The original sample studied by Sanger's team was re-sequenced in the late 1990s and is now the standard reference sequence used for comparisons. It is called the revised Cambridge Reference Sequence (rCRS). A person's mtDNA haplotype is assigned based on differences observed compared to the rCRS in the section(s) of mtDNA that are sequenced. Table 7.2 shows an example of assigning haplotypes (rather than genotypes).

The top row of sequence in Table 7.2 shows the nucleotides present at specific locations in the rCRS. The numbers refer to the location assigned in the HV region in terms of the base pair number. For example, at location 16091, the first base position, the reference and the sample both have an A (adenine) base. The sample differs from the reference in having haplotypes of 16093C (third from left) and 16129A (second to last) as defined by mutations at these positions (bold italic format in the table).

mtDNA Analysis

The analysis for mtDNA begins with extraction from the bone, hair, or other matrix. Cleaning outer surfaces is critical; mtDNA is more plentiful than nuclear DNA, so contamination is a significant concern. The sample is ground up and then the mtDNA is extracted. Next, primers isolate portions of the HV regions, followed by PCR amplification. The amplified products are then separated and sequenced using electrophoresis instruments and dye labels. Newer methods have been developed using next-generation sequencing methods that are discussed in the next chapter.

In contrast with autosomal STR analysis, which measures only a single labeled strand of DNA, both amplified mtDNA strands are usually sequenced for verification. The haplotypes are then determined by comparison to the rCRS, as shown in Table 7.2. The length of DNA sequenced is typically a few hundred base pairs long. The questioned sample (Q) result is compared to the known (K) or reference sample and data reported in terms of the number of differences in the sequences. The example in Table 7.2, for instance, shows two differences between the test sample and the reference sequence.

Heteroplasmy

One of the interesting aspects of mtDNA is that a person can have more than one type. Current research suggests that all of us have some degree of *heteroplasmy*, although at such low levels that it is not detected in a typical mtDNA analysis. There are two types of heteroplasmy – variations in sequence length and variations in the base pair sequence. For example, a person might have a mix of mtDNA molecules with CCCCCCC and CCCCCCCC (i.e., length heteroplasmy with 7 or 8 Cs), or a mix of mtDNA molecules with A and T at the same location (i.e., sequence heteroplasmy). Adding to this complication, the ratio of these variants can differ across samples and tissues. Multiple samples should be analyzed if possible. In the case of hair, several hairs would be tested when available.

Interpretation of mtDNA Data

The types of statistics and match probabilities used with STR profiles do not apply to mtDNA. Instead, analysts report results as one of three conclusions:

- Exclusion: The Q and K sequences differ at two or more locations
- Cannot exclude: The sequences have the same bases at each position or a common length variation
- Inconclusive

Suppose that the comparison of Q and K samples results in identical sequences, leading to a *cannot exclude* finding. The meaning of this is quite different than in the case of STR profiles. A match in mtDNA means that the two people have the same maternal mtDNA, but so does everyone else in the same maternal line. That line could stretch back generations. This situation can be advantageous with historical cases or when nuclear DNA is unavailable, as long as reference samples from maternal relatives are accessible. However, mtDNA provides poor discrimination power compared to the STR DNA profiles discussed in previous chapters.

mtDNA Databases and Counting

In addition to forensic databases, there are freely available mtDNA databases that are constantly expanding. The most widely used forensic population comparison is the European DNA Profiling Group (EDNAP) mtDNA Population Database (EMPOP, https://empop.online). As of July 2021, EMPOP contained almost 50,000 high-quality sequences of portions of the control region. The calculations for determining the probability of a match are different than those used with STRs. A counting method is used rather than a random match probability based on the product rule. Simplified, the probability of a sequence matching one in EMPOP corresponds to the number of times the same sequence occurs in the EMPOP database.

For example, assume the sequence identified in a case is queried against the EMPOP database and searched across racial groups, with results as shown in Table 7.3. The frequency estimate for the specific case sequence using the complete set of profiles in this population database is calculated by:

$$p \; = \; \frac{920}{50000} \; = \; 0.019 \;\; or \;\; 1.9\%$$

The same process is used to establish the initial values by racial group. As with STRs, this frequency is an estimate, and the counting method allows us to assign uncertainties using statistical methods. Most of the sequences in the database are shared by 1% or fewer of the population. Compare this to STR profiles that generate random match probabilities of one in several millions or less (see Chapter 4, Table 4.4).

One of the most interesting non-forensic applications of mtDNA is the study of human migrations. If not for mutations, our mtDNA would be identical to that of the earliest human female of our species. However, because mutations occur randomly across generations, separated human

Racial group	# Sequences in database	# Matching sequences	Match probability	% Matching
African	10,000	87	0.009	0.9
White	22,900	692	0.030	3.02
Hispanic	7,500	61	0.008	0.8
Asian	9,500	80	0.008	0.8
Across all groups	50,000	920	0.019	1.9

Table 7.3   Calculations using hypothetical results, as discussed in the text

groups have developed distinct mtDNA haplotypes over time. We can also obtain mtDNA sequences from bones and teeth, which means haplotypes can be obtained from ancient remains. This allows for the construction of migration maps, as shown in Figure 7.1. The top frame identifies the common haplogroups and their geographic origin, and the bottom frame shows the migration patterns. The date ranges are reported in years before the present and are estimated based on the known mutation rates. The origin is in Africa, with mtDNA haplogroup L0 dating back approximately 170,000 years. The most recent branch is haplogroup A, shown entering North America approximately 10,000 years ago. Haplogroup names are somewhat arbitrary but generally based on alphabetical letters, with A being the first described.

Identification of the Last Russian Tsar

An early application of mtDNA technology to a historical case was the identification of the remains of the Russian royal family killed by the Bolsheviks in 1918. Tsar Nicholas II, his wife, Alexandra, and their children (four daughters and one son) were shot and stabbed to death, along with their doctor and three servants. The reported fate of the remains was as convoluted as it was horrific. The bodies were piled into a truck that later got stuck. The bodies were then dumped in a nearby mine shaft after being stripped and soaked in acid. Finding the shaft to be too shallow for adequate concealment, the killers applied grenades to try to collapse the mine shaft. The attempt failed, and the bodies were removed and relocated to a roadside. The remains

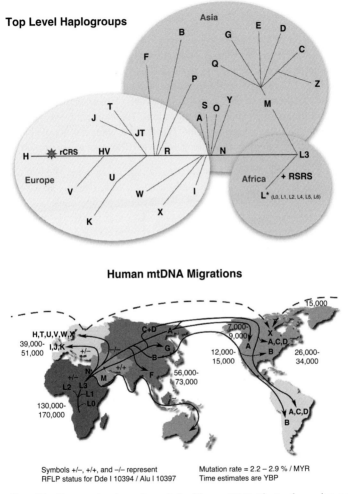

Figure 7.1   Human migration patterns derived from mtDNA. The top frame shows data by continent. The location of the Cambridge Reference Sequence is in the European circle. The origin is in Africa with the L groups, which split as humans migrated into Asia and Europe. Bottom: Groups and migrations shown on a world map. The numbers are estimates of years before the present.

were smashed with rifle butts and covered in lime. The bodies of two of the children were taken to a separate location and burned before burial.

The first site, the roadside, was rediscovered in 1979, but it was not until 1991 that action was taken. The Russian Federation requested the assistance of the UK's Forensic Science Service to evaluate the recovered remains. Testing involved five STR markers, and the results differentiated the parents' remains and three children from the doctor and servants. However, the data was insufficient to provide definitive identifications. Therefore, mtDNA was utilized. Given the known lineage of the Tsar and his wife, it was possible to obtain reference samples from living relatives. The results are shown in Figure 7.2.

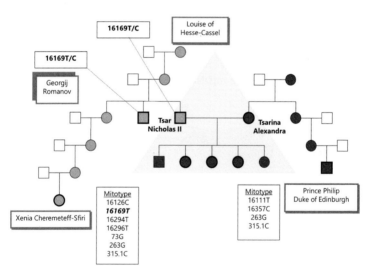

Figure 7.2   Family tree of the Russian royal family. Females are symbolized by circles and males by squares. Prince Philip (who died in 2021) was in the maternal line of the Tsarina, and Xenia Cheremeteff-Sfiri in that of the Tsar. These individuals provided reference samples. The mtDNA mitotypes (haplotypes) for each lineage appear in the boxes, with the different haplotypes indicated by different hatching patterns. The heteroplasmy is also shown.

The Tsarina is in the maternal line that includes Prince Philip, the Duke of Edinburgh, husband of Queen Elizabeth. The sample he provided had the same mtDNA haplotypes (mitotype) as the female remains recovered. The types are shown and use the terminology illustrated in Table 7.2. The samples collected from the remains thought to be the Tsar had a heteroplasmy at location 16169 (bases T and C). Reference blood from Xenia Cheremeteff-Sfiri lacked this heteroplasmy. This finding led to the exhumation of the remains of the Tsar's brother. The samples obtained from his remains had the same heteroplasmy, which was considered sufficient for definitive identification of the Tsar.

The final chapter occurred in 2007 when a smaller gravesite was located less than 100 meters from the original. Fragments of bones and teeth were recovered and indicated that the grave contained two young people, one male and one female. Testing established that the grave had held the remains of the Tsar's only son and one of his daughters. The remains were given a state funeral.

Y-Chromosome Markers

We briefly discussed the X and Y chromosomes in Chapter 4 in the context of the amelogenin (AMEL) marker used for sex identification. The Y chromosome is found in the nucleus, and it is the second smallest of the human chromosomes (Figure 4.1); the smallest is chromosome 22. The naming convention goes from largest (chromosome 1) to smallest. The Y chromosome is a paternal lineage marker, just as mtDNA is a maternal lineage marker. However, since the cell nucleus will have only a single copy of the Y chromosome, it is not as abundant as the other STRs and mitochondrial DNA. Sometimes sections of the Y chromosome are missing. In those rare situations where the AMEL-Y region is missing (AMEL-Y negative), the AMEL results will only show the AMEL-X allele when typed using the methods described in Chapter 4. This issue could lead to misidentification of a sample as being from a female rather than a male. Some STR kit manufacturers have addressed this problem by adding primers for Y-STR loci. Thus, if a sample is from a male, it will be clear from these results.

The non-recombining region of the Y chromosome (NRY), the region of the chromosome that is specific to the male, is passed down the paternal line unchanged, except for mutations, as a Y chromosome haplotype. As with mtDNA, a Y haplotype is insufficient for definitive identification as it reflects an inherited lineage. Two types of genetic markers can be examined – STR loci, using the same techniques we discussed in the Chapter 4, and single nucleotide polymorphisms (SNPs). We focus here on Y-STRs; we discuss SNPs in the next chapter.

In addition to its uses in historical studies and migration, Y-STRs can be invaluable in sexual assault cases. Sexual assault evidence may consist of DNA mixtures that require differential extraction to separate the male contribution from the female. It is not always successful, particularly if the male fraction is small. For example, it can occur when the male has had a vasectomy and no sperm cells are deposited. Mixtures of blood or blood and saliva can also be problematic, since differential extractions are not applicable. Y-STRs may also be of use in cases with more than one male contributor.

Companies have developed kits for the Y-STR markers applied in the same way as the STR kits we discussed in Chapter 4. The number of loci targeted varies, but eight or more loci are commonly used. The number of alleles found at each Y-STR locus ranges from four to more than 20. The largest publicly available database is YHRD (https://yhrd.org), which contains more than 300,000 haplotypes broken down by countries and populations. As with mtDNA, match probabilities are calculated using the counting method.

Mutation rates are important considerations with lineage markers. The reported mutation rate of the DYS458 locus is estimated as 0.66% at the YHRD site, which is one of the fastest rates for commonly used Y-STRs. This number means that there are about seven mutations per thousand generational events. A generational event is a description of the number of passages of a chromosome to the next generation. For example, if a thousand father–son pairs were examined with a mutation rate of 0.7%, we would expect about seven mutations in this dataset. Alternatively, if it were possible to explore a thousand generations in a single lineage, we would expect to see seven mutations over time in this lineage.

Table 7.4 shows some commonly used Y-STR loci and their mutation rates. Knowing these rates provides context during interpretation, particularly when minor differences exist between a questioned (Q) and a known (K) sample.

Locus (marker name)	Mutation rate	Rate as a percentage	An example haplotype
DYS19	42 in 19,807	0.21	14
DYS389I	44 in 18,051	0.24	13
DYS389II	84 in 18,104	0.46	29
DYS390	38 in 18,685	0.20	24
DYS391	46 in 19,199	0.24	11
DYS392	10 in 19,127	0.05	13
DYS393	22 in 17,977	0.12	13
DYS385	95 in 33,252	0.29	11,15

Source: Y-Chromosome STR Haplotype Reference Database, https://yhrd.org/pages/resources/mutation_rates.

Table 7.4   Mutation rates of selected Y-STRs and an example haplotype

The YHRD site also calculates the frequency of the haplotypes. For the example Y-STR haplotype shown in Table 7.4, a search of YHRD in April 2022 found 1214 matches out of 343,932 (or 1 in 283) in the database. Accounting for uncertainty, this corresponds to a range of 1 in 268 to 1 in 300. You might have noticed that DYS385 has two alleles listed, which doesn't make sense for a haplotype. This pattern arises because portions of the Y chromosome are duplicated. The DYS385 primers bind to two sections of the Y chromosome that are about 40,000 bp apart. STR alleles can be amplified from both regions. If they have different numbers of repeats (as is the case here), then two distinguishable alleles are amplified.

Thomas Jefferson and Sally Hemings

An interesting application of Y-STRs was an analysis of the paternity of a child (a son, Eston) of Sally Hemings, a slave of Thomas Jefferson. Jefferson was the third US President and author of the Declaration of Independence. He was also a lifelong slave owner. The accusation that Jefferson had fathered Hemings' son was first made in 1802 and remained a controversy for over 200 years. In the late 1990s, men in the same paternal line as Jefferson provided DNA samples, as did a direct-line male descendant from the paternal line of Eston Hemings. Y-STRs (and other markers) were characterized and found to be the same across all the samples; the haplotypes matched. Although the haplotype represents about 1% of the male population, the findings are not definitive proof that Thomas Jefferson fathered the child. The results are strong evidence that a male in the Jefferson paternal line was the father of Eston Hemings, but the data does not support a definitive identification, since Thomas Jefferson's brother or a male cousin could also have contributed the same Jefferson Y chromosome.

This case illustrates the limitations of Y-STRs for human identification; they are lineage markers with significantly less discrimination power than autosomal STRs such as described in Chapter 4. However, in a historical case such as this, traditional STR profiles are of limited use for two reasons. First, due to recombination, STR types are shuffled with every generation and thus cannot provide a continuous link across many generations. Second, because the locations of some graves are unknown or families have not given permission to exhume the remains, there is no means to obtain a viable nuclear DNA sample from Eston Hemings, Jefferson, or any of his contemporaries to perform a traditional DNA parentage test.

X-Chromosome STRs

Genetic markers on the X chromosome may also be exploited in forensic DNA profiling, although their inheritance patterns are not as straightforward as those of mtDNA. Women inherit one X from their mother and one from their father, while men inherit their X chromosome from their mother. The STRs on the X chromosome have more discrimination power than mtDNA, but the interpretation is more complex because of how they are inherited. A mother is XX, and only one of the two chromosomes passes to her children. A daughter receives one X chromosome from her mother and one X chromosome from her father. A son receives his X chromosome from his mother and his Y chromosome from his father. Children retain half the genetic information from the mother's X chromosomes, grandchildren retain a quarter, and so on through generations. This characteristic can be exploited for paternity testing and in forensic settings. X-STRs are of value in mixtures in which the male fraction is much larger than the female fraction. However, this is the reverse of typical sexual assault cases, where the male fraction may be much smaller than the female contribution.

The methods used to develop the X-STRs are like those used for autosomal STRs and Y-STRs. Kits are designed containing primers and dyes, as we have previously discussed. Primers locate the repeating units, and amplification by PCR follows. As an example, the locus DXS10101 has a repeating unit of AAAG, with the number of repeats varying between 24 and 38. Samples are analyzed using automated instrumentation, and the results are interpreted like traditional STRs. However, producing a probability statistic is more challenging with X-STRs because some of the markers may be linked, and thus their inheritance is not entirely independent from each other. As of April 2022 there are no large databases for X-STRs similar to the YHRD site described for Y-STRs.

Combined Use of Autosomal STRs and Lineage Markers

Data from lineage markers can be combined with traditional autosomal DNA profiles in a process referred to informally as *familial searching*. We discuss this topic in greater detail in Chapter 9. The first successful familial search in Australia, reported in 2019, illustrates how the combined data is utilized. A series of sexual assaults occurred in 2012 in a suburb of Adelaide. In one, a woman walking home was assaulted, and in another, a woman was able to escape before she was sexually assaulted. The first attack led to a male DNA profile that did not match any in the national database at the time. The male profile from the second attack matched the first, linking the two crimes. Police suspected many other assaults were attributed to the same assailant.

After the investigation stalled, a familial search was performed using specialized software that provides likelihood ratios (Chapter 5) to characterize partial matches. Briefly, a familial search looks for partial matches in a national STR database. The assumption is that close relatives such as siblings will share some STR genotypes since they are children of the same parents. Similarly, parents will share STR genotypes with their children. The more distant the relationship, the fewer STR genotypes in common.

The familial search in this case assisted investigators in generating a list of 100 potential sources. Next, Y-STRs were used to eliminate any of these persons of interest whose markers were not in the same paternal line as found in the crime-scene samples. Only one of the 100 had the same Y-STR haplotype as was found in the crime-scene sample. This finding was strong evidence that the perpetrator was a father or son of the person who was the source of the profile found in the familial search. This information generated a new list of POIs, which was quickly narrowed down to one man. A sample was obtained and characterized using autosomal STRs and Y-STRs, which matched the crime-scene sample. The man was convicted and sentenced to 12 years in prison.

Chapter Summary

Forensic DNA analysis has grown to include lineage markers that are inherited from mothers and fathers directly rather than from both as with the STRs described in previous chapters. These markers are useful in cases where direct comparison samples are difficult to obtain, such as in missing persons or cases in which POIs have not been identified. Autosomal STRs remain the primary method of DNA profiling, but lineage markers are increasingly being used to supplement them. Lineage markers are also useful in historical and archaeological cases because they are associated with paternal and maternal lines that persist across generations. In the next chapter we will explore new technologies and techniques that are being applied to traditional STRs as well as lineage markers.

8 Emerging Technologies

DNA profiling of STR loci is a mature technology. Improvements continue in sensitivity and additional STR loci, but the process, kits, and instrumentation are established. Courts, police, and the legal system accept and rely on DNA evidence, and databases continue to grow. However, this does not mean that the field has become static. Research continues, and newer concepts are being evaluated and adopted by the forensic community. Some of these are evolutionary, while others could be revolutionary.

SNPs

So far, we have emphasized DNA typing methods that target repeating units in STRs. The exception was mitochondrial DNA (mtDNA), where we talked about changes in a single nucleotide in a base pair sequence. Variation at a position in the mtDNA is referred to as a *single nucleotide polymorphism* or *SNP* (often pronounced "snip"). Table 7.2 in the last chapter showed examples – a C replacing a T and an A replacing a G. The number of SNPs in the human genome is estimated in the tens of millions. Most SNP loci have two alleles, but some have three. As with the STRs, most SNPs occur in non-coding regions, but some in the coding region are exploited forensically and in consumer genealogical DNA testing. A SNP in a coding region can have significant consequences in protein structure and function.

One SNP locus has less discriminating power than an STR locus because there are typically only two alleles for the SNP. However, the small size of a SNP (one base) is an advantage with degraded samples. In general, the longer the segment of bases under examination, the more susceptible it is to degradation. Low discrimination power for one SNP is countered by typing multiple SNP loci.

SNPs have several forensic applications. The first is individual identification (called *identity informative SNPs* or *IISNPs*), which requires multiple loci for good discrimination power. SNPs are also used forensically and by consumer DNA typing companies for ancestry research (*ancestry informative SNPs* or *AISNPs*). Companies target SNPs that, when taken together, are supposed to tell you about what part of the world your ancestors came from (*biogeographical ancestry*). These databases have also been used investigatively: the discovery of the Golden State Killer in 2018, as described in the next chapter, is one example. Finally, SNPs in the coding region have been exploited to reveal probable physical features such as eye color. These SNPs are called *phenotypic informative* (*PISNPs*) because they relate to physical descriptors arising from the genotype. Remember, the genotype is the actual pair of alleles from the parents, while the phenotype is how the genotype is expressed. PISNPs occur in the coding region of DNA since they impact protein synthesis and physical characteristics. More to come on this topic in Chapter 9.

King Richard III

The identification of remains unearthed in Leicester, England, illustrates how SNPs, along with lineage markers and forensic anthropology, effectively combine to identify degraded remains and long-lost missing persons. King Richard III died in battle in 1485 at the age of 32. He was thought to have been buried near a church, but the exact location was unknown. A skeleton was discovered in 2012 near the presumed site of the church, which had been torn down in the sixteenth century. Because of the potential historical importance of the find, several disciplines combined in determining whether the remains could be those of the King.

Analysis showed the skeleton to be that of a man in his early thirties with features consistent with contemporary descriptions of King Richard. Unhealed battle injuries were evident, and dating studies were consistent with the year of his death. One of the advantages in cases of royalty, which we saw with the Tsar and his family in the last chapter, is that extensive genealogical records exist and reference samples for lineage markers are often available from living relatives. This was the case with Richard, which allowed DNA studies.

The mtDNA sequence recovered from the bones matched a known living relative at every site. Interestingly, the Y-STR haplotype did not match men in the paternal line. Analysis of PISNPs recovered from the bones revealed that the man probably had blue eyes and blonde hair. This finding was consistent with an early portrait of the King. The study's authors (see King *et al.* in the references for this chapter) applied statistical methods and estimated a likelihood ratio of seven million under the assumption that the remains were from the king versus an unknown, unrelated male. This was sufficient proof of identification to rebury the remains in a royal tomb in the nearby Leicester Cathedral.

DNA Sequencing

Identifying a SNP site requires that the DNA be sequenced, meaning that nucleotides must be identified one by one in a specific order. This task requires primers and amplification, although the process is more complex than with STRs. PCR amplifies many short segments of DNA, and then computer algorithms assemble these segments into longer ones. Until recently, this task was prohibitively complex, expensive, and impractical for forensic applications.

The human genome was first sequenced by the *Human Genome Project* (*HGP*), an international project spanning 1990 to 2003. This extraordinary accomplishment mapped the approximately 3 billion base pairs in the human genome. The project involved multiple research groups and centers across the world. The complex process involved breaking the chromosomes into manageable sizes, sequencing them, and correctly ordering the sequence. DNA from several individuals was sequenced for the project. The effort required almost 15 years and cost about 3 billion US dollars.

The technique employed by the HGP involved breaking the DNA into smaller sections that were sequenced multiple times by different groups and laboratories. Since then, sequencing methods have advanced through several generations and have become available to forensic laboratories. What cost 3 billion dollars and nearly 15 years, now costs about $700 and can be done in a day by one instrument.

The current method is called *massively parallel sequencing (MPS)* or *next-generation sequencing (NGS)*. NGS instruments are now available with forensic test kits. The instrumental systems utilize different methods for parallel sequencing that do not rely on capillary electrophoresis. Companies sell kits that examine different targets and loci, including mtDNA and X/Y chromosomes, as with STR profiling.

For the identification of a mutation that results in a SNP, a reference sequence is needed. The revised Cambridge Reference Sequence (rCRS) serves this purpose for mitochondrial DNA, as outlined in Chapter 7. The standard human sequence is the Genome Reference Consortium GRCh38 human genome. This is not one person's genome sequence but a compilation from many people. The sequence is periodically updated and freely available.

SNP Case Example

NGS is already being used to help identify missing persons. Because sequencing allows investigators to look at single nucleotides, it is ideally suited for degraded DNA in which STR loci may be damaged or unusable. Sequencing is typically utilized in conjunction with other methods such as STR and mtDNA profiling. A cold case from Sweden provides an example, including the legal and privacy concerns associated with this type of case. In October 2004, an eight-year-old boy was savagely attacked and stabbed to death on a public street. A 56-year-old woman who witnessed the crime was also killed. DNA analysis revealed several mixture stains containing contributions from the assailant and the two victims. Although the killer left DNA evidence at the scene, investigators could not link the profile to any database profiles. A DNA dragnet of 6000 men failed to produce any leads, and the case went cold although the investigation continued.

Swedish authorities selected this cold case to develop policies and procedures regarding sequencing and the use of genealogy databases and websites. Investigators had at their disposal the STR profiles supplemented with sequencing analysis of mtDNA and Y-SNPs. A witness stated that the attacker appeared to be Swedish, and the lineage marker testing indicated that the man was of European descent. This hypothesis was supported by physical evidence: a cap found near the scene contained one blonde hair. The next step was whole-genome sequencing. As you might imagine, the project was labor-intensive and relied on software for analysis and interpretation. A technique called *genetic imputation* was employed. By knowing how closely linked two loci are, it is possible to predict what alleles are likely to be present at loci where data was missing.

After working through legal and privacy recommendations, selected information was submitted to two commercial genealogy sites. Several potential distant relatives were identified, and investigation led to two persons of interest, who were brothers. The STR profile of one matched the profile found on evidence recovered from the scene. He confessed and was later convicted. This case illustrates the promise and pitfalls of forensic genealogy and familial DNA, covered in the next chapter.

Sequencing STRs

The advent of NGS has allowed researchers to routinely sequence the bases associated with the STR loci used in DNA profiling (Chapter 4). Repeats of a nucleotide sequence define an STR locus. The genotype is the number of repeats from the mother and the father, such as 6,8. See Figure 4.4 to refresh your memory. This figure shows the flanking regions surrounding the repeats targeted by the primer. Sequencing allows for characterization of the nucleotides in the flanking regions as well as the STR.

Let us use the STR locus D3S1358 to show what sequencing reveals. This locus is the same one used in Figure 4.6, which shows the allelic ladder used to identify the number of repeats; here 16,17. Sequencing reveals the nucleotide sequence of the flanking and STR regions. The repeats have two units (a compound repeat), which are written as TCTA [TCTG]$_a$ [TCTA]$_b$ with the repeated sequences in brackets. The sequence of this locus and flanking regions is summarized as:

T G C C ... T C T A T C T G T C T A ... T C ... ...
1 99   176 ...   199 Flanking region
              100 175   Repeated sequences

The number below the nucleotide indicates its position within the sequence. In this example, there is one repeat of the TCTG sequence, so it is written as

Allele size range	Nucleotide sequence	Frequency
11–20	TCTA $[TCTG]_2$ $[TCTA]_n$	56%
14–20	TCTA $[TCTG]_3$ $[TCTA]_n$	28%
12–18	TCTA TCTG $[TCTA]_n$	15%

Table 8.1   Allele structures of D3S1358

TCTA $[TCTG]_1$ $[TCTA]_n$. Sequencing studies of this locus have shown the structure of the common alleles, as shown in Table 8.1. Overlap of allele size ranges means that there is more than one way to obtain some of the alleles.

Let's return to our example of a person whose type at locus D3S1358 is 16,17, as shown in Figure 4.6. These alleles could have any one of the three generic structures. Allele 16 could be structured as TCTA $[TCTG]_2$ $[TCTA]_{13}$, TCTA $[TCTG]_3$ $[TCTA]_{12}$, or TCTA TCTG $[TCTA]_{14}$. Remember, we cannot tell the specific sequence of the 16 allele based on the electropherogram (Figure 4.6). All three D3S1358 alleles would exhibit one peak at the same position (the same overall DNA fragment size) in the electropherogram. One of the advantages of sequencing is the increase in discrimination power that comes with identifying more alleles. The one 16 allele in the above example could be any one of three possibilities. In effect, sequencing has increased the number of known alleles at D3S1358 from 16 to 34. Traditional STR procedures using electrophoresis can detect 16, and sequencing can detect all 34.

Sequencing reveals SNP sites in the flanking regions, labeled as the 1–99 and 176–199 for D3S1358 (see previous page). Many STR loci, including D3S1358, do not have any SNPs with greater than 5% frequency in the population. These values may change as more samples are sequenced. Other loci show SNPs that are more polymorphic and thus potentially more useful.

Linkage and Microhaplotypes

We discussed haplotypes with Y-STRs and mtDNA in Chapter 7. *Microhaplotypes* are short sequences (less than 300 base pairs long) found in nuclear DNA that contain at least two linked SNPs. Linkage arises when sites are so close together that they tend to be inherited as a package rather than independently. This characteristic of closely spaced SNPs allows microhaplotypes to be used as lineage markers as well as for identification. Their small size is also advantageous for degraded samples. Because these sites are close to each other, they are inherited as a group and reveal ancestry information. This application is currently the most common.

Why Aren't All Labs Using NGS?

Since sequencing reveals many more alleles at STR loci, plus poly-morphic variations in the flanking regions, you might be wondering why NGS has not replaced the traditional STR method utilizing PCR and capillary electrophoresis, discussed in Chapter 4. There are several practical reasons. First, sequencing can be applied to the coding areas and thus to genes (as with PISNPs), which introduces privacy concerns. Second, sample preparation differs from PCR methods and is more labor-intensive. Interpretation of the results requires greater computational power, and storing massive data files is costly. Technology and termin-ology are still being developed, and the community has yet to agree on a core set of SNPs analogous to the STR loci. Core loci are critical for kit development and validation, which means that more laboratories will use them, and more allele frequency data will be generated. If only a few labs are typing a set of SNPs, population frequencies will be based on small datasets, making them less likely to represent larger groups. None of these issues are insurmountable, but it will take time for large-scale implemen-tation and application of NGS.

Rapid DNA

A significant technological advancement for STR profiling is the development of *rapid DNA* instruments. These devices are integrated systems that generate STR profiles in about 90 minutes. They are designed for single-source samples such as cheek swabs collected at police stations or immigration checkpoints. Some agencies deploy rapid DNA systems to crime scenes and mass disaster sites as well. As of July 2021, the FBI has authorized accredited laboratories and police booking stations in the US to upload profiles obtained on selected instruments into the CODIS database. This authorization does not extend to other applications or operations. For example, a crime-scene sample run on a rapid DNA instrument cannot be uploaded to the national DNA database because it might be a complicated mixture and not amenable to an automated interpretation.

Rapid DNA employs the same process as traditional STR profiling with extraction, PCR amplification, and electrophoretic separation. The difference is that all functions are miniaturized and integrated into a single disposable cartridge. Consequently, the cost of obtaining a profile is higher compared to laboratory methods. Field operation requires some training but not the same extensive education and training background required for laboratory analysts. It will take a few years to see how rapid DNA systems are employed and to assess their value to the forensic and law enforcement communities.

One area in which rapid DNA has already made inroads is in mass fatality incidents. An example was identification of remains after the devasting Camp Fire in November 2018. The fire spread quickly and burned at a ferocious pace approaching 80 acres (32 hectares) per minute. Eighty-five people were killed in and around the town of Paradise, making it the deadliest wildfire in California's history. The condition of the remains complicated identification of the deceased. Samples were usually burned beyond recognition and consisted of fragmentary bones, tissue, and teeth. In mass casualty events such as this fire, a small local DNA database can be developed without uploading DNA profiles to criminal databases, making it an ideal application of rapid DNA technology. The response also illustrates how different disciplines are integrated into identifying remains, similar to what we saw with King Richard III. In this case, teams consisted of volunteer searchers, anthropologists and anthropology students, pathologists, and dentists.

Once the area was safe, teams searched through damaged and destroyed structures and flagged locations where there appeared to be human remains. Recovery teams documented the sites, including personal items that might be useful in identification. Remains were transported to a morgue with a rapid DNA system housed in a mobile laboratory in the parking lot. Anthropologists examined bones and classified them by sex and approximate age if the conditions of the remains allowed. Dental records were also employed, as were fingerprints. A few cases were resolved based on medical devices and implants recovered with the remains, such as knee and shoulder implants. These protocols worked for 22 of the 85 cases, with the remainder (69 sets of remains) characterized by rapid DNA. Reference DNA samples were collected from relatives of known or potential victims. Sample types included bone, dried blood, clotted blood, swabs of organ tissue, and muscle. In about 90% of the cases, a viable STR profile was obtained and subjected to familial comparisons. Rapid DNA identified 58 people.

Non-Human DNA

Forensic application of non-human DNA has grown in the past few years in ways that may surprise you. Among the first extensions of DNA profiling from humans was a 1994 murder case in Canada. A woman disappeared from her home, and her car with a bloodstained interior was located a few days later. A few weeks later, a bloody jacket was recovered, and the DNA matched the victim. White cat hairs were found on the lining. Eventually, the victim's body was located in a shallow grave, and her common-law husband was identified as a person of interest. The man was living with his parents at the time, who owned a white cat named Snowball. An STR profile from 10 loci was obtained from Snowball's hair and blood. The random match probabilities were in the range of one in a billion. The man was later convicted of second-degree murder. This case is an example of DNA as trace evidence; the transfer of the cat hair to the jacket helped link that jacket to the suspect.

The close relationship of humans with pets drove many of the early non-human DNA applications, and STRs for cats, dogs, horses, and cattle are available online (https://strbase.nist.gov). As with human DNA methods, domestic animal DNA applications include mtDNA, SNPs, and haplotypes. You may have used a commercial kit to test your dog or cat; breed information is based on SNPs and microhaplotypes. Standard sequences and databases analogous to those we have discussed for humans have been developed and deployed over the years.

Case Examples

A case from Argentina illustrated how dog mtDNA could link a suspect to a scene. The crime was a robbery and homicide, and the critical evidence was dog feces. Investigators identified a POI and collected his shoes as evidence, since fecal matter was embedded in crevices in the sole. The resident dog provided reference samples. As you might predict, dog feces are a challenging matrix that requires extensive sample preparation to remove contaminants that can interfere with PCR amplification.

Analysts attempted analysis of 15 STR loci without success, but mtDNA sequencing worked. Data analysis led to the calculation of a likelihood ratio of about 20, meaning the evidence obtained was 20 times more likely if the dog feces recovered from the shoe were from the dog that lived where the murder occurred than a randomly selected dog.

In a case from 2016, a hunting dog was accidentally shot and killed during a wild boar hunt. During emergency surgery, wild boar hair was found in the dog's wound. The organizers of the hunting event kept track of which hunters shot which boars, so hair from all boars was collected. STR typing of the hair produced a match between the hair recovered from the wound and one of the boars killed during the hunt, implicating that hunter in the accidental shooting.

The role of DNA in animal forensics is expanding. Animal forensics can be divided into domestic animals, such as pets and farm animals, and wildlife. Wildlife forensics has become a recognized discipline, and several laboratories are devoted exclusively to its practice. Examples include the Wildlife DNA Forensics Laboratory in Scotland and the US Fish and Wildlife facility in Ashland, Oregon. Wildlife forensics are integrated into the investigation of crimes such as hunting endangered species, trafficking animal products (pelts, ivory, horns, etc.), poisoning, and poaching. The same types of DNA analysis used for human DNA (STRs, mtDNA, and SNPs) have been applied to animals and plants. However, for many wildlife forensic applications, commercial kits are not available.

DNA and Marijuana

DNA techniques have been adapted to the analysis of cannabis plants and plant material. Two categories of cannabis are of interest – hemp (used for fiber and other consumer products) and marijuana (used for the intoxicating effects of tetrahydrocannabinol, THC). In the US, as one example, THC is a controlled substance at the federal level, while many states have legalized or decriminalized its use. At the same time, consumers have embraced related compounds such as cannabidiol (CBD) as a home remedy for ailments such as insomnia and arthritis pain. Hemp plants produce low levels of THC but high levels of CBD. The mixed market and legal implications mean that there is a need to distinguish hemp plants from marijuana plants. Current research centers on sequencing two genes found in both plant types. The genes code for enzymes that are integral to the synthesis of THC and CBD. Sequencing studies have uncovered SNP sites that differ between hemp and marijuana and appear promising as a rapid means of distinguishing them.

Microbial Forensics

The anthrax attacks in the US following the terrorist strikes of September 11, 2001 introduced the world to microbial forensic science. The FBI was concerned on that day that one or more of the highjacked airplanes might contain a biological weapon. This was not the case, but the first sign of a genuine biological attack surfaced in Florida three weeks later. Two men who worked in a mailroom arrived at Miami-area hospitals with symptoms including delirium, nausea, and vomiting. The men continued to decline despite antibiotic treatment. About two days later, a spinal fluid test of one revealed *Bacillus anthracis* or anthrax. Anthrax is a naturally occurring bacterium found in soil, which complicated the identification. One of the men died soon after, but the other eventually recovered.

The spinal fluid sample was tested by two laboratories using VNTR DNA technologies (recall this was early in the DNA era). The strain was identified as Ames, which was used in research. This finding meant that a biological weapon had been deployed. Subsequent mailings of envelopes containing powders to national news networks and members of Congress confirmed suspicions. Eventually, five people died, and 17 became ill. The last fatality came in late November 2001.

Scientists working at a government laboratory set up cultures to increase the number of available samples. This turned out to be key to the case. Several of the resulting colonies appeared slightly different from others. Researchers thought these variants (seven) were from DNA mutations. An original Ames strain sample from 1981 was located and used to generate a reference sequence for comparison. Researchers from several laboratories participated in the project to sequence the variants and interpret the results. FBI agents scoured the US and the world for laboratories that had the Ames strain and collected samples. PCR methods were adopted as part of the sequencing, speeding the tedious work. SNP sites were identified and utilized for differentiation. The effort required several years of intensive investigation and laboratory work. Eventually, eight of 1082 bacterial samples were linked to a single source held by one government microbiologist, Bruce Ivins. He had been immunized against anthrax before the attacks began. He committed suicide in 2008 and the case was closed.

Sequencing methods today are much faster and have opened many avenues of microbial forensics. Examples include profiles of bacteria on the skin, soil bacteria, and gut bacteria. Food- and waterborne diseases are also targets. Viruses are included within microbial forensics, but here the focus is on RNA since viruses do not have DNA. Protocols and methods are similar and rely on sequencing.

Forensic Epigenetics

DNA segments that provide the code for building specific proteins represent a small portion of our DNA. Non-coding regions can still play a role in how genes are expressed and how our environment influences our genes. This field of study is called *epigenetics*. Simplified, epigenetic changes play a role in determining when genes are activated and when they are shut down. The epigenetic change of most forensic interest is methylation, which means a chemical unit called a methyl group is added to DNA. Methylation is driven by factors such as age, environment, and behaviors. Three areas of most interest are identification of body fluid type (blood, menstrual blood, and vaginal fluid); age estimation; and differentiation of identical twins. While forensic epigenetics are not commonly used in casework, some assays have been developed and applied to forensic samples.

Developing tests for epigenetic markers typically involves identifying which genes (and their DNA sequences) to target and from what tissues. Testing methods are then developed to measure the degree of methylation. Once the genes to be targeted are identified, the next step is to gather a large and representative dataset from individuals. The degree of methylation at the targeted loci can then be linked to the characteristic to generate a comparative scale.

As more genetic markers are identified through medical research, more characteristics will be amenable to forensic epigenetic methods. However, there are ethical and privacy concerns that must be addressed. Unlike STR methods, epigenetics target coding regions (genes) and reveal personal and medical information about the donor. This potential impact on privacy will be central in the next and final chapter.

Chapter Summary

This chapter has provided a glimpse into the possible future of forensic DNA analysis. Advances such as NGS have provided new capabilities and possibilities that will define the broader field of forensic genetics (and epigenetics). Sequencing STRs has revealed more alleles and variants than are detected with current methods. Forensic DNA testing is not limited to humans, as was illustrated in the case of the anthrax attacks in 2001. Finally, rapid DNA technologies have allowed some STR typing to be performed outside the traditional laboratory environment. Inevitably, these emerging technologies have led to new challenges and questions related to privacy and security of information, among other topics. The next chapter highlights some of these emerging issues in forensic DNA analysis and how they have impacted cases, policies, and legislation.

9 Emerging Issues

The relentless advance of DNA typing capabilities leads to complications and concerns. It is one thing for someone to be able to obtain your ABO blood type from a tiny spot of blood and another thing to know your eye color and ancestry. Some of the concerns are due to misconceptions, but others can pit personal privacy against perceived security. Practices and policies have not caught up with capability. We highlight a few of these current dilemmas in this chapter.

Interpretation of Complex DNA Mixtures

We have discussed simple mixtures and the challenges presented by complex mixtures containing DNA from more than two contributors. An excellent website to refresh your memory is the *DNA Mixtures: A Forensic Science Explainer* site (see the references for this chapter). Complexity arises from DNA transfer and trace DNA, degradation, and stochastic effects. Manual interpretation becomes more challenging with multiple contributors, and software is used to model peak height variation and probabilistically account for stochastic effects in predicting possible genotype combinations of contributors. There is always a concern when trust is transferred from human interpretation to an algorithm in a machine. Testing, evaluation, and validation have become more critical than ever before. Mixture interpretation has become one of the most discussed topics in current DNA typing.

Mixture Interpretation and Cognitive Bias

Many research studies in the past decade have shown how mixture interpretations can vary, be they obtained with or without *probabilistic genotyping software (PGS)*, which we first mentioned in Chapter 5. Many factors contribute to differences in data interpretation; one might surprise you. There is a misconception that DNA evidence is entirely neutral and objective, but this is not the case. As with any human endeavor, subjectivity is involved. Analysts make choices as part of mixture interpretation even with software. Software cannot be completely objective because humans design the algorithms, develop the software code, input the data, and select the settings. Subjectivity can be minimized by testing, review, and other procedures, but it is never eliminated. DNA evidence is no exception.

A study published in 2011 starkly revealed patterns of subjectivity in DNA interpretation (see Dror and Hampikian in the references). The authors obtained raw data from a DNA mixture involving multiple assailants in an actual sexual assault case. One of the suspects cooperated with investigators and identified others he said were involved in the attack. His cooperation was part of a plea bargain offering a lesser sentence. The men he implicated denied involvement. Under state law, the testimony of the admitted rapist would not be allowed unless supported by evidence. The DNA mixture analysis became critical. If the mixture analysis excluded the men or was inconclusive, the admitted rapist probably would not have been allowed to testify against the men he implicated and therefore could not have benefited from the plea deal. The laboratory reported that one of the four suspects could not be excluded. The DNA analysts in the actual case knew of the admissibility restriction. Did the fact that the analysts knew about the restriction influence how they interpreted the analytical data? The 2011 study was designed to provide insight into this type of situation, and to see if such prior knowledge might introduce subjectivity and bias to mixture interpretation.

Information critical to the case but not to the DNA analysis is called *domain irrelevant context*. Here, the domain irrelevant context is the information regarding the testimony and potential plea deal of the one admitted assailant. In a purely objective mixture analysis, knowing the consequences of the result would not influence the mixture interpretation. To evaluate the impact of domain irrelevant context, the study's authors presented the raw data without the contextual information to seventeen DNA experts who knew nothing of the case, trial, or outcome. All worked at the same government laboratory and thus operated under the same policies, procedures, and interpretation guidelines. It would be reasonable to expect that all the analysts would produce the same interpretation since they had the same data and worked under the same policies and guidelines. This is not what happened. Of the 17 analysts, only one reached the same conclusion as in the actual case. Four reported inconclusive findings and twelve reported an exclusion.

Although this was a small study, it yielded two key findings. First, since only one of the 17 analysts agreed with the original analyst, this suggests that the irrelevant knowledge might have influenced the initial interpretation in the actual case. Second, the study illustrates how frequently DNA analysts come to different conclusions about the same mixture data. This finding is the core of the issue with mixtures. These 17 analysts, working in the same laboratory under the same guidelines, reached different conclusions with the same data and information. Subsequent studies reiterated the need to improve mixture interpretation. Software packages have been developed for this purpose, but this has not to date proven to be a complete solution.

Proprietary Software

Law enforcement agencies increasingly utilize software in advanced applications such as latent fingerprint analysis, facial recognition, and, of interest here, PGS used to interpret DNA mixtures. Briefly, the data obtained from the electropherogram is analyzed by the PGS software to generate the mixture interpretation. The information generated by the probabilistic genotyping software is referred to as *algorithmic evidence*. The need for such software packages is clear, and they have become integral to the justice system. However, even the most advanced software comes with limitations that must be considered and addressed in implementation. Reliability is the key issue.

We introduced PGS in Chapter 5. We learned how challenging mixture analysis can be and how software can enable examination of a wider range of DNA mixtures than manual interpretation approaches. Companies responded to the need by developing software that evaluates mixtures and assigns likelihood ratios (LRs). These commercial products took years of effort to develop, so, understandably, companies are reluctant to release their programs' source codes. We noted a similar situation in the early days of STR profiling with primers (Chapter 4). Manufacturers were reluctant to publish the sequence of these primers since they are the basis of commercial kits purchased by forensic laboratories. Eventually, companies were compelled to provide primer data for court proceedings. The defendant has the right to know how STR profiles are developed and ensure that the procedures are trustworthy.

Without looking at the source code, it is impossible to know how a program makes decisions and what potential problems might exist. This doesn't mean the software is faulty or untrustworthy, but it does require thorough validation. As an example, if you use Excel, you assume that calculations are correct. This trust is earned because of extensive validation and testing. Open-source software is an alternative to commercial products, but these packages are also complex and may not have the same infrastructure for support and updates as a commercial product. Currently, commercial and open-source PGS packages are used in casework.

Laboratories validate software before implementation. The validation process involves testing the software under different conditions using different samples. The software's results are compared to human interpretations using synthetic control materials and past case samples. Validation is a detailed and tedious process but it is essential to establish reliability. Validation does not mean the system is infallible, but rather indicates that a laboratory conducted experiments to study a method's capabilities and limitations and decided that the method was fit for their purpose.

Reliability

A brief history of PGS illustrates how the concept of reliability has evolved and what challenges remain. By the late 1990s, the concepts of probabilistic genotyping were being proposed, and early computer programs were developed in the first decade of the twenty-first century. Around this same time, DNA methods had become sensitive enough to apply to trace and touch DNA, influencing the types of evidence submitted to laboratories. As capabilities improved, so did the demand for low-level mixture analysis. This need pushed software development. The laudable goals were to ensure accuracy, reduce subjectivity, and reduce inconsistencies in interpretation.

A landmark report, published in 2016 by the President's Council of Advisors on Science and Technology (PCAST), addressed PGS and highlighted the promise as well as issues of concern (pp. 78–79, emphasis added):

> These probabilistic genotyping software programs clearly represent a major improvement over purely subjective interpretation. However, they still require careful scrutiny to determine (1) whether the methods are scientifically valid, including defining the limitations on their reliability (that is, the circumstances in which they may yield unreliable results) and (2) whether the software correctly implements the methods. This is particularly important because the programs employ different mathematical algorithms and *can yield different results for the same mixture profile*.
>
> Appropriate evaluation of the proposed methods should consist of studies by multiple groups, *not associated with the software developers*, that investigate the performance and define the limitations of programs by testing them on a wide range of mixtures with different properties. Three entities are involved in ensuring the reliability of commercial PGS. First is the company that sells it, second is the laboratory that purchases it, and third are independent entities. The latter are essential to ensure objectivity.

The report further noted that the performance of software applied to mixtures of more than three contributors had not been adequately studied or validated. A supplement published in 2017 added that addressing such concerns should be clear-cut and involve testing on a wide range of samples.

An important observation was that different software packages could yield different results with the same data. We saw this same issue with manual interpretation of mixtures in our discussion of cognitive bias at the beginning of the chapter. Thus, we are back to the issue of inconsistency. The reports did not mention source-code access as critical; rather they emphasized the need for additional reliability testing.

Three entities are involved in the reliability testing of products and procedures used in forensic science. The first is the company that makes and sells the software. The second is the laboratory deploying the software, which must conduct an internal validation before its implementation in casework. Third are groups and associations independent of the company or laboratory. Organizations such as the UK Forensic Regulator, SWGDAM, and the International Society for Forensic Genetics have published recommendations on how forensic laboratories should approach reliability testing and validation of PGS.

Since the PCAST report was published, some extensive validation studies of PGS software have been published, and others have been made available to forensic laboratories. However, debate continues regarding how to characterize the reliability of mixture interpretation provided by software packages. There is no question that PGS has become an essential tool for mixture interpretation, but discussion surrounding validation and best practices continues.

Contradictions

Although the results of PGS have been admitted into judicial proceedings since 2009, legal challenges still arise. A central concern is that different software packages can produce different and sometimes contradictory results and LR values for the interpretation. As we have seen, the same is true of DNA analysts, but in that case, it is possible to question these analysts to understand why and how they reached their conclusions. The general algorithms used in commercial software are known, but the source code is considered a trade secret. Thus, it can be challenging to deduce why results differ and what those differences mean. Ideally, if the same electropherogram of a mixture is presented to several different analysts or different software packages, the analysts and software would reach the same conclusions and produce similar likelihood ratios. In some instances, the differences are explainable, such as they are using different population databases, or they are selecting different options in the software. In other cases, the differences remain puzzling, which leads to the obvious question: Which is the most reliable?

First, realize that there are normal expected variations within results. There is little practical difference between a random match probability of one in 3 billion and one in 10 billion. A likelihood ratio of 95 would be considered by most practitioners to convey the same information as an LR of 90. Probabilistic statements are never absolute, so minor variations are not a concern. However, the expected range of variation must be characterized.

Second, it is hard to gauge how critical it is to see the source code of a commercial software package line by line. Validation studies, which employ known control samples, can probe the performance of software across a range of conditions and sample types. The worry is that a case might represent a unique and untested situation not covered in validation studies. Would this situation be recognized as an issue? This risk can be difficult to characterize.

Contradictory results across software packages (exclusion vs. inclusion, for example) are more problematic and may come down to how the software is coded. When this situation arises with manual interpretation by analysts, these people can be questioned to provide insight into why results varied. This is a more difficult task with software. Thus, while PGS has become vital to mixture interpretation, some issues remain unsettled. That controversy remains should not surprise us; recall our discussions regarding the admissibility of DNA evidence from Chapter 3. The adoption of new technology into forensic science is rarely smooth, nor do we want it to be. Scientific and legal wrangling is inevitable – and indispensable for building the necessary understanding and trust in the methods used.

Cases and Controversy

In 2011, Florencio Jose Dominguez was convicted of murder committed in 2008. He had been tried twice, with the first trial ending in a hung jury. In the second, he was sentenced to 50 years to life in prison. The key evidence was a DNA mixture recovered from a pair of bloody gloves found at the crime scene. A PGS package provided critical mixture interpretation. The timing was everything in this case. The laboratory had updated the way it reported data from mixture analysis and, had the new method been applied to Dominguez's case, it would have resulted in inconclusive findings rather than inclusion.

However, trial testimony was based on the older method, and the change had not been reported to the defense.

Dominguez's lawyer learned of the issue in 2017 and was able to have the conviction reversed. Prosecutors moved forward to begin the third trial. The defense asked to have an expert review the source code for the PGS package, and the company agreed, with the caveat that a strict non-disclosure agreement be signed, and other criteria met. The defense countered that this was too restrictive and would violate his client's constitutional rights, a common argument in such situations. The judge told the prosecutors that he might exclude the DNA evidence from a new trial if the source code was not provided. This led to an agreement in which Dominguez pleaded guilty to manslaughter and was released in 2019 based on time served.

A similar case involving another PGS began with a 2014 gas station robbery in Virginia. One robber grabbed the attendant's shirt while the other emptied the cash drawer. DNA testing at the time was unsuccessful, but the evidence was retained. In the meantime, investigators obtained information from a man in prison that he was the one who took the money. He identified the other robber, the man who grabbed the shirt. The shirt was retested in 2019, and this time a mixture was found. The PGS was used to generate LRs compared to the newly identified suspect. The LR implicating the new suspect was in the quadrillions. The laboratory using the software had conducted validation and reliability studies. The defense argued that this was insufficient and that the source code should be reviewed. They argued that errors had been found in other packages when such a review was undertaken.

The PGS developer responded to the request with requirements similar to those in the Dominguez case. These included that the expert obtain liability insurance, travel to the company site, and take handwritten notes. The cost of these requirements was not trivial. It meant buying insurance, travel, and many hours of work to accomplish a review. The defense would bear these costs. Additionally, the conditions of the non-disclosure agreement were such that the expert might not have been able to testify at trial, since these proceedings are part of the public record. The defense argued that the cost was prohibitive and asked to exclude the mixture results at trial. The judge required that the PGS developer remove all restrictions but agreed that the code should not be released. While admissibility challenges continue to be raised in US courts, PGS results on DNA mixtures have been used in several hundred thousand cases in the past few years.

Indirect DNA Searching, Familial DNA, and Kinship Analysis

DNA is now being used to find persons of interest through their relatives. If a profile submitted to a database does not produce a match, indirect approaches such as identifying relatives can be employed. Several tools are available. First, familial searching methods use law enforcement databases and partial matches to find possible close relatives such as siblings or parents/children. The second method is Y-STR database searches, as described in Chapter 7, revealing information on paternal lineage. Paternal data information can be combined with mtDNA for maternal lineage information. The third method is investigative genetic genealogy (IGG). We will tackle this topic in a later section in the chapter. All these methods provide investigative information designed to lead to a person of interest and a direct reference sample being collected for evaluation purposes.

Familial Searching

Familial searching is the process of searching law enforcement databases to identify close biological relatives (first-degree relatives such as siblings, parents, and children). The idea is that first-degree relatives share a significant portion of DNA, around 50%, including alleles from STR markers. Therefore, partial matches may indicate a biological relationship between the individual contributing the evidence sample and the individual behind the DNA database profile. However, unrelated individuals may share some DNA characteristics, including specific common STR alleles. In the US, familial searching often involves additional screening of potentially related database samples with Y-STR markers, since law enforcement DNA databases typically contain more male than female profiles. Many offenders have relatives whose profiles are in the arrestee or offender indices; this is an underlying assumption behind partial-match searches. The Y-STR screening can help separate those samples which randomly possess alleles shared with the evidence profile versus a father, brother, son, uncle, or other close paternal relative who is biologically related.

The UK recognized the potential value of partial matches and familial searches and began using the technique in 2002. The first conviction came in a 2004 manslaughter case in which the assailant threw a brick off a bridge. It went through a truck window, hit the driver in the chest, and caused a fatal heart attack. The DNA data helped to find the perpetrator, who admitted the crime and was convicted of manslaughter. The first Australian conviction was obtained in 2019 and related to rapes committed in 2015. We discussed this case in Chapter 7.

French police first utilized familial searching in a 2002 rape/murder. A woman involved in a car crash called emergency services. She was heard screaming while two other voices were heard in the background. The car was found the next morning, and her partially burned body was recovered from a dumpster the following day. Semen was recovered as part of the evidence. The DNA profile covered 18 loci, but no database matches were found. Mass DNA typing of men in the region (5300 individuals) did not provide any leads. Additional searches over the years were unsuccessful.

The familial database search was conducted in 2011. One male was identified with at least one allele in common with the crime-scene evidence at all 18 loci. Common male lineage of this individual with the crime-scene sample was confirmed using Y-STRs. The identified man was a direct male relative of the man who left the semen at the scene. This finding was the critical investigative lead needed. The man lived in the region, so the next step was to find his father and other sons. Based on their ages at the time of the crime, the father and one of the sons were eliminated. The older son, who was 23 at the time, died a short time after the crime, which explained how he escaped the DNA dragnet. The police obtained a DNA sample from his mother rather than exhume the body, which investigators feared would warn any accomplices. This DNA analysis showed that the dead son was the source of the crime-scene DNA.

The Grim Sleeper serial killer case in the US utilized familial DNA searching. A serial killer active in the 1980s in Los Angeles killed at least 10 women. This was before DNA typing became routine. In 2008, a DNA profile from crime-scene evidence was searched against the California DNA database with no matches produced. A familial search was unsuccessful. However, a second familial search of the California DNA data in 2010 identified a potential first-degree relative. This individual's DNA had been collected as the result of a weapons charge after the first search was conducted. Follow-up Y-STR testing matched this individual from the database with the Grim Sleeper case evidence. Investigators identified this individual's father as a person of interest and surreptitiously collected a discarded piece of pizza from the older man. The profile matched that of the killer, and Lonnie Franklin Jr. was arrested in 2010. He was convicted in 2016 and died in prison in 2020.

Investigative Genetic Genealogy (IGG)

Before we dive into forensic genealogy, we need to review DNA databases and what they contain. DNA profiles from STR loci are stored in law enforcement databases. Different countries have different policies regarding the entry of data, but the public cannot access the databases. The DNA profiles stored in these databases do not contain any phenotype information, because STRs are taken from non-coding DNA regions.

Commercial companies now provide DNA testing kits direct to consumers (DTC). A person purchases a kit and sends a sample such as saliva or a buccal swab to the company. The sample is analyzed, and the results are stored in the company's database. These kits do not type STR loci used in the law enforcement database. Instead, they examine large numbers of single nucleotide polymorphism (SNP) markers to explore ancestry and in some cases genetic diseases.

These companies use microarray chips to characterize SNPs, typically 650,000–700,000 sites. It is essential to understand that this analysis does not overlap with typical forensic laboratory DNA profiling of STR loci. STRs are used for identification, while commercial SNP typing is designed for genealogical purposes. Forensic laboratories can conduct Y- and X-STR profiling using commercial kits, and some have access to SNP analysis. However, there is typically no overlap between the genetic markers used in forensic and consumer DNA databases. Some companies offer additional services such as mtDNA, Y-STRs, and genetic testing for specific health conditions. By purchasing a testing kit and having a profile generated, you can access database services such as ancestry analysis. Some companies allow law enforcement to upload profiles for searching while others do not, although law enforcement entities can seek and have sought access. This is an area in which policies and procedures are evolving to catch up with the technology. In the US, one famous case brought these issues into the public spotlight.

The Golden State Killer

The case of the Golden State Killer exemplifies the promise and potential perils of IGG. The techniques were successful in catching a violent rapist and serial killer from California. Still, they involved searches of commercial databases that in some cases contradicted privacy policies and expectations of users. This case and others like it have sparked debate pitting individual privacy against a potentially valuable law enforcement technique. In turn, the discussion is leading to new policies, procedures, and recommendations.

The known crimes began in 1976 with a rape in Rancho Cordova, California, near the state capital Sacramento. Numerous rapes and attempted rapes continued in the general area and were thought to be the work of the same assailant. This was well before the DNA era, so linking crimes was not straightforward. The perpetrator acquired the nickname of the East Area Rapist, among others. The first known murders occurred in February 1978 when a couple were shot while walking their dog. In 1978, the crimes moved to the San Francisco Bay area. The rapes and murders spread to southern California in 1980. There was a five-year pause in attacks from 1981 to 1986, with the last known attack occurring in 1986. Physical evidence was collected in all the cases, but much of the evidence pertaining to the sexual assaults was destroyed when the statute of limitations on the rapes expired. The killer's spree included at least 12 murders, 50 rapes, and more than 100 burglaries.

The investigation by local, state, and federal law enforcement continued over the years. DNA methods were applied as they became available, but the perpetrator's STR profile was not found in CODIS. In 2001, the Sacramento cases were linked to the crimes in the Bay area and southern California with STR testing. The FBI released additional information and offered a reward in 2016, which returned the case to public attention.

In 2017, a stored rape kit was used to generate an autosomal DNA SNP file that could be compared against profiles in genetic genealogy databases. An initial search at a free genealogy site was unsuccessful. Eventually, investigators submitted the profile to two other DTC vendors. These searches identified possible distant cousins. In early 2018, a genealogy consultant working with the FBI uploaded the SNP profile to a third genealogy site. The consultant, Barbara Rae Venter, was able with genealogical record sleuthing to develop a family tree of the perpetrator. After a few weeks, Venter narrowed the list to two possible men. Surreptitious sampling and traditional STR DNA testing eliminated one and found a matching profile for the other.

As a result, Joseph James DeAngelo was arrested in April 2018 when he was 72 years old. He had been a police officer in the region where the attacks occurred from 1973 to 1979. He was fired in 1979 for shoplifting. It was impossible to link him to all the crimes, because the statute of limitations had expired on the rapes. He was charged with multiple counts of murder and kidnapping. He pleaded guilty in 2020 and was sentenced to multiple life sentences.

The Controversy

The IGG practices used in the Golden State Killer case were effective but raised ethical questions focusing on user privacy and expectations. Commercial DNA testing and their large databases were designed to reveal ancestry information and to connect users with their biological relatives. If you have submitted a kit for testing, you are likely interested in learning more about your family history. If this data is accessed by law enforcement, this not only involves your information but that of your relatives who may not have submitted kits. As a result, your relatives might be subject to surveillance and surreptitious sample collection for DNA profiling. Yet these procedures led investigators to the perpetrator of a horrible series of crimes. The challenge is balancing the needs of public safety and personal genetic privacy.

In the Golden State Killer case, one DTC vendor worked with law enforcement while two others were unaware of how their sites and data were utilized. These and other companies have since adjusted their privacy policies and developed procedures for law enforcement requests. Not surprisingly, shortly after the Golden State Killer story became known, companies began offering IGG services. Government entities entered the discussion by publishing guidelines and recommendations. For example, the US Department of Justice published interim guidelines in 2019 that emphasized that IGG should only be used after traditional DNA profiling with STRs has been undertaken and results submitted to CODIS. The term "STRs first and last" was coined, meaning that an STR profile must be developed and searched before moving on to IGG. IGG is intended to create investigative information that will eventually lead to persons of interest and another STR profile for confirmation purposes. SWGDAM published an overview of IGG in 2020 that reiterated the concept of "STR first and last," which is another way of reinforcing that IGG is a tool for developing investigative leads rather than identifying individuals.

DTC companies have modified their policies such that the potential for use by law enforcement entities is clear, which provides informed consent to consumers. If you submit a kit for testing, you can download your results, but other consumers cannot. Users can also remove their data if they wish. Health and medical data are not used as part of IGG. Thus, it appears that IGG has matured quickly, and policies and procedures have developed that provide reasonable protections. In 2021, Maryland became the first US state to regulate IGG activities.

Forensic DNA Phenotyping

STRs and most SNPs that we have discussed so far are from non-coding regions of DNA. The exception is phenotypic informative SNPs (PISNPs), which we touched on in the last chapter. Having DNA sequence information has facilitated the ability to predict physical characteristics and features from biological evidence. This capability is also implicit in DTC genetic testing, which predicts ancestry and produces results that can sometimes be linked to likely physical characteristics such as skin color. *Forensic DNA phenotyping* (*FDP*) zeros in on loci known to influence features such as skin color, eye color, and hair color. As with IGG, phenotyping aims to generate investigative leads in criminal and missing person cases. The case of King Richard III from the previous chapter highlighted SNPs and physical appearance.

Features targeted in FDP fall into two categories: *externally visible characteristics* (*EVP*) and *biogeographical ancestry* (*BGA*). Information is gleaned from selected SNPs that are known to impact the characteristic. Note that features such as eye color can be predicted with reasonable but not absolute certainty. A tool targeting SNPs related to eye color called IrisPlex was introduced in 2011, and has since been expanded to include hair color predictions via HIrisPlex (https://hirisplex.erasmusmc.nl). The website is a joint project of multiple academic institutions and is available to all. For example, the IrisPlex system utilizes six SNPs within specific genes to predict eye color, while the combined hair, skin, and eye color webtool HIrisPlex-S uses 41 SNP markers. Accuracy metrics reported at the site range from about 0.5 (50%) to 0.9 (90%). The algorithm is least effective for intermediate eye colors like hazel or green (no better than chance at 50%) and most effective in predicting blue eyes and dark to black skin color (90%).

FDP is expanding into other characteristics as information regarding the relevant genes becomes available. One example is hair loss and baldness, in which 12 genes have been identified; another is hair type such as curly or straight. The Golden State Killer case involved a prediction of baldness that turned out to be correct. One company has become well known for offering FDP that includes facial appearance predictions. This company provides a product that generates a facial rendering that resembles a composite artist's sketch to accompany FDP information. The efficacy of this method for obtaining reasonable likenesses appears to have been mixed to date. The method faces several challenges, including not only cognitive bias but also alteration of externally visible characteristics through surgery, hormone therapy, or temporary means. On the other hand, as a lead-generating method in cold cases, FDP could be very helpful.

FDP and Missing Persons

The DNA Doe Project (https://dnadoeproject.org) utilizes IGG and phenotyping to identify unidentified human remains (UHR). Volunteers operate the project and work with provided data. Recently, the project successfully identified one of the few remaining unidentified victims of the Green River Killer.

Gary Ridgway killed at least 49 women in the Seattle area in the 1980s. He was finally caught in 2001. The unidentified woman's skeletal remains were found in a swampy area near the Seattle–Tacoma airport in 1984. Ridgway admitted to killing her and dumping the body, but he did not know who she was.

Bone fragments were submitted to a laboratory to generate genetic data (SNPs used for genealogical research). The bones indicated that she was very young, and it turned out that she was the youngest of the known victims. The results were forwarded to the DNA Doe Project, which uploaded the profile to two DTC sites. Several distant cousins were found but no closer relatives. The next step was the construction of many family trees to find common ancestors. Public records and other sources of genealogical research were employed. Volunteers identified the girl's parents. Ironically, the girl's mother had uploaded her own DNA into one of the DTC sites in early 2019 in hopes she could find her daughter.

The reason the mother's profile was not immediately flagged was a policy change at the DTC company. In the aftermath of the Golden State Killer case, genealogical companies and sites changed privacy policies to reflect potential use by law enforcement. The site in question adopted an opt-in policy that meant users had to specifically allow law enforcement access to their information. The mother was not aware of the change and had not opted in. All the work paid off when the unknown victim was identified as Wendy Stephens, 14, who had run away from home in Denver in 1983.

Chapter Summary

Mixture interpretation has become a pressing challenge in DNA profiling, now that the methods are capable of detecting trace levels of DNA. Probabilistic genotyping programs developed to help interpret mixtures have been a boon to forensic laboratories, but their inherent complexity has led to debates and controversy within the scientific and legal communities. As with human interpretation, the core issues are reliability and consistency. Breakthroughs in forensic DNA phenotyping have assisted in missing persons cases but raise other issues. Finally, we have seen how investigative genetic genealogy (IGG) has rapidly become a valuable and unique tool for generating investigative leads, particularly in cold cases. IGG is unique in that it involves direct-to-consumer DNA databases. However, the final DNA analysis in these cases always comes back to STR DNA profiling, which will likely remain the primary DNA typing tool for years to come.

Concluding Remarks

We have come a long way in a short time. From the first use of DNA typing for a criminal investigation in 1986 to now, over 35 years have passed. Those years have brought a revolutionary change in human identification, from ABO blood typing to analysis of complex mixtures, probabilistic genotyping software, and investigative genetic genealogy. Forensic DNA typing now applies to STRs (still the primary method), Y-STRs, mitochondrial DNA, and SNPs. We have law enforcement databases and consumer databases that are used in current and cold cases. We have seen how portable DNA instruments can be used in mass fatalities and police booking stations.

There are several key takeaways from this journey, including the need to correct several common misunderstandings, as summarized in the next section. Perhaps the most critical point is that finding someone's DNA on a piece of evidence tells you nothing about how, when, or why it got there. Only a thorough investigation can hope to address these questions. Also, as we have seen, DNA evidence is powerful but not infallible. The limitations of any forensic method must always be weighed when it is used.

The evolution and acceptance of DNA methods have fundamentally changed forensic science. Implementing probabilistic evidence such as the random match probability and likelihood ratios in DNA has pushed other forensic disciplines to adopt statistical approaches. Progress has been uneven and, in some cases, contentious, but progress it is. The impact of DNA methods in forensic science is hard to overstate, and it will be fascinating to see where it goes in the next 35 years.

Summary of Common Misunderstandings

DNA evidence is infallible. DNA evidence is a powerful investigative tool, but it is not infallible. The limitations of any forensic method must always be considered when using it. See Chapters 2 and 6.

A DNA profile is unique to an individual and provides definitive identification of that person. We don't know whether any profile is unique. Population frequencies help estimate the chance of a randomly selected person having the same genotype as the one of interest. Still, the result is always expressed as a probability, not a certainty. See Chapter 1.

DNA profiles focus on genes. Routine DNA typing uses short tandem repeats (STRs) from non-coding DNA. The only DNA profiling that does target genes is associated with selected single nucleotide polymorphisms (SNPs). See Chapters 1 and 8.

Science and justice are natural partners. These disciplines are founded on different principles and have different goals. The justice system exists to settle disputes, while science strives to understand the natural world. See Chapter 3.

DNA used in profiling is junk DNA. Traditional STR profiling targets non-coding regions, and the functions of this DNA are the subject of much research. However, the label "junk" does not apply. See Chapter 3.

DNA evidence was quickly embraced by the legal system. It took several years from the mid-1980s through the mid-1990s for most legal challenges to the methodology to be addressed. See Chapter 3.

DNA evidence solves the case. DNA evidence is part of larger investigations. It may provide breakthroughs in the case, but it alone never solves it. See Chapter 3.

DNA profiling can provide information about your physical appearance, health, or other genetic traits. Traditional STR typing targets non-coding DNA that does not directly determine physical characteristics. See Chapter 4.

The random match probability is the probability that a DNA profile came from a specific person. This probability expresses the chance that someone has the same STR profile as another person. It *does not* mean that there is only one person in the population with this type; there may be more. We cannot know the exact frequency; allele frequencies used to calculate the random match probability provide an estimate based on the individuals typed in a specific set of population samples. Probabilities are based on data in frequency databases. Unless every human being is typed, we can never know whether a profile came from a specific person and not from any other. See Chapter 4.

The random match probability can be used to estimate the probability of guilt or innocence. It should never be used this way. It can be used to evaluate the weight of a matching profile, but this information must be integrated into other results from the investigation. See Chapter 4.

DNA databases contain the entire genomic sequence. Law enforcement DNA databases contain only STR profiles. See Chapter 5.

DNA databases contain names. Law enforcement databases contain STR profiles. Identities are stored separately. See Chapter 5.

CODIS is a single national database in the US. CODIS (the Combined DNA Index System) is the software used to connect a system of linked local and state databases and indices for the National DNA Index System (NDIS), which is managed by the FBI. See Chapter 5.

There is one giant worldwide DNA database. Different countries have their own databases. Joint searches are possible if the STR profiles use the same loci and data formats are consistent. See Chapter 5.

DNA evidence is always probative. DNA is found everywhere. Only DNA deposited during a crime is relevant to that crime. Finding DNA doesn't tell you why it is there, how it got there, when it got there, and what it means. See Chapters 6 and 9.

Your DNA comes from both your father and your mother. Your mitochondrial DNA comes from your mother and her maternal line. In males, the Y-chromosome DNA comes from the father and his paternal line. See Chapter 7. STRs used in DNA profiling are inherited from mothers and fathers, but other DNA is not. See Chapters 1 and 4.

References and Further Reading

Chapter 1

American Red Cross (2021). Facts about blood and blood types. www .redcrossblood.org/donate-blood/blood-types.html (accessed August 12, 2021).

Ballantyne, J. (2000). Serology. In *Encyclopedia of Forensic Sciences*, ed. J. A. Seigel. San Diego, CA: Academic Press/Elsevier.

Chapter 2

American Red Cross (2021). Facts about blood and blood types. www .redcrossblood.org/donate-blood/blood-types.html (accessed August 12, 2021).

Ballantyne, J. (2000). Serology. In *Encyclopedia of Forensic Sciences*, ed. J. A. Seigel. San Diego, CA: Academic Press/Elsevier.

Bromwich, M. R. (2007). *Final Report of the Independent Investigator for the Houston Police Department Crime Laboratory and Property Room*. Washington, DC: Fried, Frank, Harris, Shriver & Jacobson LLP.

Gaensslen, R. E., Bell, S. C., and Lee, H. C. (1987). Distributions of genetic markers in United States populations. 1. Blood group and secretor systems. *Journal of Forensic Sciences* **32**(4): 1016–1058.

Gaensslen, R. E., Bell, S. C., and Lee, H. C. (1987). Distributions of genetic markers in United States populations. 2. Isoenzyme systems. *Journal of Forensic Sciences* **32**(5): 1348–1381.

Mitra, R., Mishra, N., and Rath, G. P. (2014). Blood groups systems. *Indian Journal of Anaesthesia* **58**(5): 524–528.

Wang, L., Chen, M., Wang, F., *et al.* (2020). A 21-plex system of STRs integrated with Y-STR DYS391 and ABO typing for forensic DNA analysis. *Australian Journal of Forensic Sciences* **52**(1): 16–26.

WorldAtlas (2021). World population by percentage of blood types. www.worldatlas .com/articles/what-are-the-different-blood-types.html (accessed August 18, 2021).

Wraxall, B. G. D., and Culliford, B. J. (1968). A thin-layer starch gel method for enzyme typing of bloodstains. *Journal of the Forensic Science Society* 8(2–3): 81–82.

Chapter 3

Gill, P., and Werrett, D. J. (1987). Exclusion of a man charged with murder by DNA fingerprinting. *Forensic Science International* **35**(2–3): 145–148.

National Institute of Justice (2000). *The Future of Forensic DNA Typing: Predictions of the Research and Development Working Group*. Washington, DC: US Department of Justice.

National Research Council (1992). *DNA Technology in Forensic Science*. Washington, DC: National Academies Press. https://doi.org/10.17226/1866.

National Research Council (1996). *The Evaluation of Forensic DNA Evidence*. Washington, DC: National Academies Press. https://doi.org/10.17226/5141.

University of Leicester (2009). The history of genetic fingerprinting. https://le.ac.uk /dna-fingerprinting/history (accessed August 15, 2021).

Venter, C. H. (2020). A critical review of the current state of forensic science knowledge and its integration in legal systems. PhD thesis, University of South Africa.

Wagner, J. K. (2013). Out with the "junk DNA" phrase. *Journal of Forensic Sciences* 58(1): 292–294.

Zagorski, N. (2006). Profile of Alec J. Jeffreys. *Proceedings of the National Academy of Sciences* 103(24): 8918–8920.

Chapter 4

Gasiorowski-Denis, E. (2016). The mystery of the Phantom of Heilbronn. *ISOFOCUS* July/August 2016. www.iso.org/news/2016/07/Ref2094.html (accessed August 18, 2021).

Ogawa, H., Hiroshige, Y., Yoshimoto, T., Ishii, A., and Yamamoto, T. (2018). STR genotyping from a dry-cleaned skirt in a sexual assault case. *Journal of Forensic Sciences* **63**(4): 1291-1297.

Royal Society of Edinburgh (2017). *Forensic DNA Analysis: A Primer for Courts*. Edinburgh: RSE. https://royalsociety.org/~/media/about-us/programmes/sci ence-and-law/royal-society-forensic-dna-analysis-primer-for-courts.pdf (accessed August 19, 2021).

Scientific Working Group on DNA Analysis Methods (2020). SWGDAM Quality Assurance Standards Documents, 2020 Quality Assurance Standards. www .swgdam.org/publications (accessed August 19, 2021).

Chapter 5

Bieber, F. R., Buckleton, J. S., Budowle, B., Butler, J. M., and Coble, M. D. (2016). Evaluation of forensic DNA mixture evidence: protocol for evaluation, interpretation, and statistical calculations using the combined probability of inclusion. *BMC Genetics* **17**(1): 125. https://doi.org/10.1186/s12863-016-04 29-7.

Graversen, T., Mortera, J., and Lago, G. (2019). The Yara Gambirasio case: combining evidence in a complex DNA mixture case. *Forensic Science International: Genetics* **40**: 52–63.

Criminal Justice Information Services. Frequently asked questions on CODIS and NDIS. www.fbi.gov/services/laboratory/biometric-analysis/codis/codis-and-ndis-fact-sheet (accessed August 19, 2021).

Royal Society of Edinburgh (2017). *Forensic DNA Analysis: A Primer for Courts*. Edinburgh: RSE. https://royalsociety.org/~/media/about-us/programmes/sci ence-and-law/royal-society-forensic-dna-analysis-primer-for-courts.pdf (accessed August 19, 2021).

Scientific Working Group on DNA Analysis Methods (2019). *Recommendations of the SWGDAM Ad Hoc Working Group on Genotyping Results Reported as Likelihood Ratios*. SWGDAM. www.swgdam.org/publications (accessed August 19, 2021)

Torres, Y., Flores, I., Prieto, V., *et al.* (2003). DNA mixtures in forensic casework: a 4-year retrospective study. *Forensic Science International* **134**(2–3): 180–186.

Xiao, C., Jiang, Y. W., and Liang, M. (2019). Using the information embedded in the mixed profiles to assist in determining the identity of the deceased and the suspect in a deficiency case. *Forensic Science International* **300**: E13–E19.

Chapter 6

Basset, P. and Castella, V. (2018). Lessons from a study of DNA contaminations from police services and forensic laboratories in Switzerland. *Forensic Science International: Genetics* **33**: 147–154.

Gill, P. (2016). Analysis and implications of the miscarriages of justice of Amanda Knox and Raffaele Sollecito. *Forensic Science International: Genetics* **23**: 9–18.

Gill, P. (2019). DNA evidence and miscarriages of justice. *Forensic Science International* **294**: E1–E3.

Gosch, A., and Courts, C. (2019). On DNA transfer: the lack and difficulty of systematic research and how to do it better. *Forensic Science International: Genetics* **40**: 24–36.

Gosch, A., Euteneuer, J., Preuss-Wössner, J., and Courts, C. (2020). DNA transfer to firearms in alternative realistic handling scenarios. *Forensic Science International: Genetics* **48**: 102355.

Kanokwongnuwut, P., Kirkbride, K. P., and Linacre, A. (2018). Detection of latent DNA. *Forensic Science International: Genetics* **37**: 95–101.

Meakin, G. E., Butcher, E. V., van Oorschot, R. A. H., and Morgan, R. M. (2017). Trace DNA evidence dynamics: an investigation into the deposition and persistence of directly- and indirectly-transferred DNA on regularly-used knives. *Forensic Science International: Genetics* **29**: 38–47.

Smith, P. A. (2016). When DNA implicates the innocent. *Scientific American* **314** (6): 11–12. https://doi.org/10.1038/scientificamerican0616-11.

Thornbury, D., Goray, M., and van Oorschot, R. A. H. (2021). Indirect DNA transfer without contact from dried biological materials on various surfaces. *Forensic Science International: Genetics* **51**: 102457.

van Oorschot, R. A. H., Szkuta, B., Meakin, G. E., Kokshoorn, B, and Goray, M. (2019). DNA transfer in forensic science: a review. *Forensic Science International: Genetics* **38**: 140–166.

Chapter 7

Abarno, D., Sobieraj, T., Summers, C., and Taylor, D. (2019). The first Australian conviction resulting from a familial search. *Australian Journal of Forensic Sciences* **51**: S56–S59.

Byard, R. W. (2020). The execution of the Romanov family at Yekatarinberg. *Forensic Science, Medicine, and Pathology* **16**(3): 552–556.

Coble, M. D., Loreille, O. M., Wadhams, M. J., et al. (2009). Mystery solved: the identification of the two missing Romanov children using DNA analysis. *PLoS One* **4**(3): e4838. https://doi.org/10.1371/journal.pone.0004838.

Davis, L. (1998). Mitochondrial DNA: State of Tennessee v. Paul Ware. *GenePrint*. www.promega.com/~/media/Files/Resources/Profiles%20In%20DNA/103/Mit ochondrial%20DNA%20State%20of%20Tennessee%20v%20Paul%20Ware .ashx (accessed August 19, 2021).

Gill, P., Ivanov,P. L., Kimpton,C., et al. (1994). Identification of the Romanov family by DNA analysis. *Nature Genetics* **6**(2): 130–135.

Ivanov, P. L., Wadhams, M. J., Roby, R. K., et al. (1996). Mitochondrial DNA sequence heteroplasmy in the Grand Duke of Russia Georgij Romanov establishes the authenticity of the remains of Tsar Nicholas II. *Nature Genetics* **12**(4): 417–420.

Marquez, M. C., Caballero, P. A. B., and Domingo, C. N. (2021). Application of mitochondrial DNA as a tool in the forensic field: update and new perspectives in mtDNA analysis. In *Forensic DNA Analysis: Technological Developments and Innovative Applications*, ed. E. Pilli and A. Berti. Boca Raton, FL: CRC Press, pp. 113–146.

Melton, T. (2016). Mitochondrial DNA: profiling. In *A Guide to Forensic DNA Profiling*, ed. S. Bader and A. Jamieson. New York: John Wiley & Sons, pp. 397–405.

NBC News (2003) Judge: DNA found in Peterson's boat admissible. NBC, November 17, 2003. www.nbcnews.com/id/wbna3474357 (accessed August 28, 2021).

Pereira, V., and Gusmao, L. (2021). The X-chromosomal STRs in forensic genetics: X chromosome STRs. In *Forensic DNA Analysis: Technological Developments and Innovative Applications*, ed. E. Pilli and A. Berti. Boca Raton, FL: CRC Press, pp. 91–112.

Royal Society of Edinburgh (2017). *Forensic DNA Analysis: A Primer for Courts*. Edinburgh: RSE. https://royalsociety.org/~/media/about-us/programmes/sci ence-and-law/royal-society-forensic-dna-analysis-primer-for-courts.pdf (accessed August 19, 2021).

Vergani, D. (2021). The Y-chromosomal STRs in forensic genetics: Y chromosome STRs. In *Forensic DNA Analysis: Technological Developments and Innovative Applications*, ed. E. Pilli and A. Berti. Boca Raton, FL: CRC Press, pp. 77–90.

Wang, L., Chen, M., Wang, F., *et al.* (2020). A 21-plex system of STRs integrated with Y-STR DYS391 and ABO typing for forensic DNA analysis. *Australian Journal of Forensic Sciences* **52**(1): 16–26.

Chapter 8

Ballard, D., Winkler-Galicki, J., and Wesoly, J. (2020). Massive parallel sequencing in forensics: advantages, issues, technicalities, and prospects. *International Journal of Legal Medicine* **134**(4): 1291–1303.

Barrientos, L. S., Crespi, J. A., Fameli, A., *et al.* (2016). DNA profile of dog feces as evidence to solve a homicide. *Legal Medicine* **22**: 54-57.

Borsting, C., Pererira, V., Andersen, J. D., and Morling, N. (2016). Single nucleotide polymorphism. In *A Guide to Forensic DNA Profiling*, ed. A. Jamieson and S. Bader. Chichester: John Wiley & Sons, pp. 205–222.

Bruijns, B., Tiggelaar, R., and Gardeniers, H. (2018). Massively parallel sequencing techniques for forensics: a review. *Electrophoresis* **39**(21): 2642–2654.

Butler, J. M., David, V. A., O'Brien, S. J., and Menotti-Raymond, M. A. (2002). The MeowPlex: a new DNA test using tetranucleotide STR markers for the domestic Cat. *Profiles in DNA* 5(2), 7–10.

Decker, R. S., and Kerns, T. L. (2020). The FBI's Amerithrax Task Force and the advent of microbial forensics. In *Microbial Forensics* (Third Edition), ed. B. Budowle, S. Schutzer and S. A. Morse. San Diego, CA: Academic Press, pp. 11–23.

de Knijff, P. (2019). From next generation sequencing to now generation sequencing in forensics. *Forensic Science International: Genetics* **38**: 175–180.

Forlani, G., and Petrollino, D. (2021). A comprehensive molecular approach to the detection of drug-type versus fiber-type hemp varieties. *Forensic Science International: Genetics* **52**: 102464.

Genome Reference Consortium. Human Build 38 patch release 14 (GRCh38.p14). www.ncbi.nlm.nih.gov/assembly/GCA_000001405.29 (accessed April 12, 2022).

Gettings, K. B., Aponte, R. A., Vallone, P. M., and Butler, J. M. (2015). STR allele sequence variation: current knowledge and future issues. *Forensic Science International: Genetics* **18**: 118–130.

Gettings, K. B., Borsuk, L. A., Steffen, C. R., Kiesler, K. M., and Vallone, P. M. (2018). Sequence-based US population data for 27 autosomal STR loci. *Forensic Science International: Genetics* **37**: 106–115.

Gin, K., Tovar, J., Bartelink, E. J., *et al.* (2020). The 2018 California wildfires: integration of rapid DNA to dramatically accelerate victim identification. *Journal of Forensic Sciences* **65**(3): 791–799.

Jackson, P. J. (2020). Microbial forensic investigation of the anthrax letter attacks: how the investigation would differ using today's technologies. In *Microbial Forensics* (Third Edition), ed. B. Budowle, S. Schutzer and S. A. Morse. San Diego, CA: Academic Press, pp. 25–32.

King, T. E., Gonzalez Fortes, G., Balaresque, P., *et al.* (2014). Identification of the remains of King Richard III. *Nature Communications* **5**: 5631.

Linacre, A. (2021). Animal forensic genetics. *Genes* **12**(4): 515.

National Human Genome Research Institute, National Institutes of Health (2020). The Human Genome Project. www.genome.gov/human-genome-project (accessed August 19, 2021).

SASA [Scottish Agricultural Science Agency]. Wildlife crime. www.sasa.gov.uk/wildlife-environment/wildlife-crime (accessed April 12, 2022).

Schleimer, A., A. C. Frantz, J. Lang, P. Reinert and M. Heddergott (2016). Identifying a hunter responsible for killing a hunting dog by individual-specific genetic profiling of wild boar DNA transferred to the canine during the accidental shooting. *Forensic Science Medicine and Pathology* **12**(4): 491–496.

Thermo Fisher Scientific (2021). FBI approves Thermo Fisher Scientific's rapid DNA solution for use in law enforcement booking stations. *Cision PR Newswire*, July 8, 2021. www.prnewswire.com/news-releases/fbi-approves-thermo-fisher-scientifics-rapid-dna-solution-for-use-in-law-enforcement-booking-stations-301327486.html (accessed August 29, 2021).

Tillmar, A., Fagerholm, S. A., Staaf, J., Sjolund, P., and Ansell, R. (2021). Getting the conclusive lead with investigative genetic genealogy: a successful case study of a 16 year old double murder in Sweden. *Forensic Science International: Genetics* **53**: 102525.

US Fish and Wildlife Service. Clark R. Bavin National Fish and Wildlife Forensics Laboratory. www.fws.gov/law-enforcement/clark-r-bavin-national-fish-and-wildlife-forensics-laboratory (accessed April 12, 2022).

Vidaki, A., and Kayser, M. (2017). From forensic epigenetics to forensic epigenomics: broadening DNA investigative intelligence. *Genome Biology* **18**(1): 238.

Vidaki, A., and Kayser, M. (2018). Recent progress, methods and perspectives in forensic epigenetics. *Forensic Science International: Genetics* **37**: 180–195.

Yamamuro, T., Segawa, H., Kuwayama, K., *et al.* (2021). Rapid identification of drug-type and fiber-type cannabis by allele specific duplex PCR. *Forensic Science International* **318**: 110634.

Chapter 9

Arnold, C. (2020). The controversial company using DNA to sketch the faces of criminals. *Nature* **585**: 178–181.

Barrio, P. A., Crespillo, M., Luque, J. A., *et al.* (2018). GHEP-ISFG collaborative exercise on mixture profiles (GHEP-MIX06). Reporting conclusions: results and evaluation. *Forensic Science International: Genetics* **35**: 156–163.

Butler, J. M., Iver, H., Press, R., *et al.* (2021). *DNA Mixture Interpretation: A NIST Scientific Foundation Review* (NISTIR 8351-DRAFT). Gaithersburg, MD: National Institute of Standards and Technology, US Department of Commerce.

Butler, J. M., Kline, M. C., and Coble, M. D. (2018). NIST interlaboratory studies involving DNA mixtures (MIX05 and MIX13): variation observed and lessons learned. *Forensic Science International: Genetics* **37**: 81–94.

Coble, M. D., and Bright, J. A. (2019). Probabilistic genotyping software: an overview. *Forensic Science International: Genetics* **38**: 219–224.

Dror, I. E., and Hampikian, G. (2011). Subjectivity and bias in forensic DNA mixture interpretation. *Science & Justice* **51**(4): 204–208.

Egel, B. (2018). Here's the string of crimes tied to the East Area Rapist in years of California terror. *The Sacramento Bee*, April 18, 2018.

Ge, J. Y., and Budowle, B. (2021). Forensic investigation approaches of searching relatives in DNA databases. *Journal of Forensic Sciences.* **66**(2): 430–443. https://doi.org/10.1111/1556-4029.14615.

Gill, P., Kirkham, A., and Curran, J. (2007). LoComatioN: a software tool for the analysis of low copy number DNA profiles. *Forensic Science International* **166** (2–3): 128–138.

Government Accountability Office (2021). Forensic technology: algorithms strengthen forensic analysis, but several factors can affect outcomes. GAO-21-435SP. Washington, DC: Government Accountability Office. www.gao.gov/products/gao-21-435sp (accessed August 19, 2021).

Greytak, E. M., Moore, C., and Armentrout, S. L. (2019). Genetic genealogy for cold case and active investigations. *Forensic Science International* **299**: 103–113.

Jouvenal, J. (2021). A secret algorithm is transforming DNA evidence. This defendant could be the first to scrutinize it. *Washington Post*, July 13, 2021.

Kamb, L. (2021). DNA puts a name to one of the last unidentified victims of the Green River Killer. *Seattle Times*, January 25, 2021.

Katsanis, S. H. (2020). Pedigrees and perpetrators: uses of DNA and genealogy in forensic investigations. *Annual Review of Genomics and Human Genetics* **21**: 535–564.

Kayser, M. (2015). Forensic DNA phenotyping: predicting human appearance from crime scene material for investigative purposes. *Forensic Science International: Genetics* **18**: 33–48.

Kennett, D. (2019). Using genetic genealogy databases in missing persons cases and to develop suspect leads in violent crimes. *Forensic Science International* **301**: 107–117.

Kling, D., Phillips, C., Kennett, D., and Tillmar, A. (2021). Investigative genetic genealogy: current methods, knowledge and practice. *Forensic Science International: Genetics* **52**: 102474.

Krimsky, S. (2022). *Understanding DNA Ancestry*. Cambridge: Cambridge University Press.

Maguire, C. N., McCallum, L. A., Storey, C., and Whitaker, J. P. (2014). Familial searching: a specialist forensic DNA profiling service utilising the National DNA Database (R) to identify unknown offenders via their relatives. The UK experience. *Forensic Science International: Genetics* **8**(1): 1–9.

Mateen, R. M., Sabar, M. F., Hussain, S., Parveen, R., and Hussain, M. (2021). Familial DNA analysis and criminal investigation: usage, downsides and privacy concerns. *Forensic Science International* **318**: 110576.

Miller, G. (2010). Familial DNA testing scores a win in serial killer case. *Science* **329**(5989): 262. https://doi.org/10.1126/science.329.5989.262.

Moran, G. (2019). Murder case that highlighted DNA analysis controversy ends with plea to reduce charge, release. *San Diego Union Tribune*, December 9, 2019.

Moran, K. S. (2018). Damned by DNA: balancing personal privacy with public safety. *Forensic Science International* **292**: E3–E4.

Mortera, J. (2020). DNA mixtures in forensic investigations: the statistical state of the art. *Annual Review of Statistics and Its Application* **7**: 111–142.

Pham-Hoai, E., Crispino, F., and Hampikian, G. (2014). The first successful use of a low stringency familial match in a French criminal investigation. *Journal of Forensic Sciences* **59**(3): 816–819.

Plemel, E. (2019). Genetic genealogy and its use in criminal investigations: are we heading towards a universal genetic database? *Dalhousie Journal of Interdisciplinary Management* **15**. https://doi.org/10.5931/djim.v15i0.8983.

President's Council of Advisors on Science and Technology (2016). *Forensic Science in the Criminal Courts: Ensuring Scientific Validity of Feature Comparison Methods*. Washington, DC: PCAST.

President's Council of Advisors on Science and Technology (2017). *Addendum to Forensic Science in the Criminal Courts: Ensuring Scientific Validity of Feature Comparison Methods*. Washington DC, PCAST.

Press, R. (2019) *DNA Mixtures: A Forensic Science Explainer*. Gaithersburg, MD: National Institute of Standards and Technology, US Department of Commerce. www.nist.gov/feature-stories/dna-mixtures-forensic-science-explainer (accessed April 8, 2022).

Samuel, G., and Prainsack, B. (2019). Civil society stakeholder views on forensic DNA phenotyping: balancing risks and benefits. *Forensic Science International: Genetics* **43**: 102157.

Scientific Working Group on DNA Analysis Methods (2020). Overview of investigative genetic genealogy. SWGDAM. www.swgdam.org/publications (accessed August 19, 2021).

Scudder, N., McNevin, D., Kelty, S. F., Walsh, S. J., and Robertson, J. (2018). Forensic DNA phenotyping: developing a model privacy impact assessment. *Forensic Science International: Genetics* **34**: 222–230.

St. John, P. (2020). The untold story of how the Golden State Killer was found. *Los Angeles Times*, December 8, 2020.

US Department of Justice (2019). *United States Department of Justice Interim Policy Forensic Genetic Genealogical DNA Analysis and Searching*. Washington, DC: US Department of Justice.

Figure Credits

Figure 1.1   Elements obtained from the US National Institutes of Health and from Figures 2.3 and 2.4, Butler, J. M. (2010). *Fundamentals of Forensic DNA Typing.* San Diego, CA: Academic Press/Elsevier.

Figure 1.2   Adapted from US government source, www.genome.gov/genetics-glossary/Nucleotide.

Figure 1.3   Adapted with permission from Figure 2.6, Butler J. M. (2010). *Fundamentals of Forensic DNA Typing.* San Diego, CA: Academic Press/Elsevier.

Figure 3.1   Some elements adapted with permission from Figure 3.4, Butler, J. M. (2010). *Fundamentals of Forensic DNA Typing.* San Diego, CA: Academic Press/Elsevier.

Figure 3.2   Reproduced with permission from Gill, P., and Werrett, D. J. (1987). Exclusion of a man charged with murder by DNA fingerprinting. *Forensic Science International* 35(2–3): 145–148. © Elsevier.

Figure 4.1   Adapted with permission from Figure 2.3, Butler, J. M. (2010). *Fundamentals of Forensic DNA Typing.* San Diego, CA: Academic Press/Elsevier.

Figure 4.2   Adapted with permission from figures in Chapter 9, Butler, J. M. (2010). *Fundamentals of Forensic DNA Typing.* San Diego, CA: Academic Press/Elsevier.

Figure 4.3   Adapted with permission from figures in Chapter 9, Butler, J. M. (2010). *Fundamentals of Forensic DNA Typing.* San Diego, CA: Academic Press/Elsevier.

Figure 4.4   Adapted with permission from PowerPoint presentation available at https://strbase.nist.gov/training.htm (*Fundamentals of Forensic DNA Typing*, Chapter 8) .

Figure 4.5   Adapted with permission from Figure 8.6, Butler J. M. (2010). *Fundamentals of Forensic DNA Typing*. San Diego, CA: Academic Press/Elsevier.

Figure 4.6   Adapted with permission from PowerPoint presentation available at https://strbase.nist.gov/training.htm (*Fundamentals of Forensic DNA Typing*, Chapter 8).

Figure 4.7   Adapted with permission from PowerPoint presentation available at https://strbase.nist.gov/training.htm (*Fundamentals of Forensic DNA Typing*, Chapter 8).

Figure 5.1   Adapted with permission from Figure 2.6, Butler, J. M. (2015). *Advanced Topics in Forensic DNA Typing: Interpretation*. San Diego, CA: Academic Press/Elsevier.

Figure 5.2   Reproduced with permission from Figure 8.2, Butler, J. M. (2012). *Advanced Topics in Forensic DNA Typing: Methodology*. San Diego, CA: Academic Press/Elsevier.

Figures 5.3, Adapted with permission from Figures 6.1 and 6.2 and Box 6.2,
5.4, and 5.5 Butler, J. M. (2015). *Advanced Topics in Forensic DNA Typing: Interpretation*. San Diego, CA: Academic Press/Elsevier.

Figure 6.1   Adapted with permission from Figures 4.5 and 11.2, Butler, J. M. (2012). *Advanced Topics in Forensic DNA Typing: Methodology*. San Diego, CA: Academic Press/Elsevier.

Figure 6.2   Reproduced with permission from Kanokwongnuwut, P., Kirkbride, K. P., and Linacre, A. (2018). Detection of latent DNA. *Forensic Science International: Genetics* **37**: 95–101. © Elsevier.

Figure 7.1   Reproduced from www.mitomap.org. Creative Commons Attribution 3.0 (CC BY 3.0).

Figure 7.2   Reproduced with permission from Box 16.5, Butler J. M. (2010). *Fundamentals of Forensic DNA Typing*. San Diego, CA: Academic Press/Elsevier.

Index

Tables are denoted in **bold**; figures in *italics*